# CONTEMPORARY MATHEMATICS

## Volume 25

# Value Distribution Theory and its Applications

### Chung-Chun Yang, Editor

AMERICAN MATHEMATICAL SOCIETY

Providence · Rhode Island

PROCEEDINGS OF THE SPECIAL SESSION
ON VALUE DISTRIBUTION THEORY AND ITS APPLICATIONS
803RD MEETING OF THE AMERICAN MATHEMATICAL SOCIETY

HELD IN NEW YORK CITY,

NEW YORK

APRIL 14–15, 1983


1980 *Mathematics Subject Classification.* Primary 30Dxx, 32Axx, 30D30, 30D35, 32A22.

**Library of Congress Cataloging in Publication Data**

Special Session on Value Distribution Theory and Its Applications (1983: New York, N. Y.)
Value distribution theory and its applications.
    (Contemporary mathematics, ISSN 0271-4132; v. 25)
    "Proceedings of the Special Session on Value Distribution Theory and Its Applications, 803rd Meeting of the American Mathematical Society, held in New York City, New York, April 14–15, 1983" — T.p. verso.
    Bibliography: p.
    1. Value distribution theory—Congresses. I. Yang, Chung-Chun, 1942– . II. American Mathematical Society. Meeting (803rd : 1983 : New York, N.Y.) III. Title. IV. Series: Contemporary mathematics (American Mathematical Society); v. 25.
QA331.S687   1983              515.7'82          83-21465
ISBN 0-8218-5025-3

**CONTEMPORARY MATHEMATICS**

## Titles in this Series

# Titles in this series

CONTENTS

PREFACE

Recently there have appeared numerous articles on the beautiful and
classical Nevanlinna value distribution theory and its applications.  More
specifically, in areas such as defect relations, growth estimation of solutions
of differential equations, distribution of zeros and poles of meromorphic
functions, entire functions of bounded index, factorization theory, etc.
These results have appeared in many different journals in various parts of the
world.

In April of 1983 a special session on value distribution theory and its
application was held in New York City as part of the 803rd meeting of the
American Mathematical Society.  The purpose of this session was to bring together some of the mathematicians working in this active field, to present the
results of their research, to explore and exchange problems among themselves
and to stimulate further research in this area.

Most of the papers in this collection were presented at the special session
and have never before been published.  For instance, announced here by Charles
Osgood is his complete proof of the validity of the analogous Nevanlinna
second fundamental theorem for small functions (i.e. defect relations for small
functions).  This has been a challenge problem for complex analysts for a
long time.

I wish to thank all the mathematicians who participated in the special
session and for their contributions to this volume.  Special thanks are extended
to those from abroad:  Professors Chi-Tai Chuang and Yong-Xing Gu from China,
Professors Kiyoshi Niino and Seiki Mori from Japan, Professor Shlomo Strelitz
from Israel, Professor Sakari Toppila from Finland, and Professor L.S.O.
Liverpool from Nigeria.  Also I wish to thank the American Mathematical Society
and the editorial board of the Contemporary Mathematics Series for their
approval of this publication and also the editorial department for many helpful
suggestions.

In conclusion, it is the unanimous feeling of all the contributors to this
volume that the late Professor Rolf Nevanlinna has inspired us in many different
ways in developing and searching for further results in this field, both in
one variable and in several variables.  Therefore it is a privilege and our
opportunity to acknowledge our debt to him and to dedicate this collection to
his memory.

CHUNG-CHUN YANG
WASHINGTON, D.C.

DEDICATED TO THE MEMORY OF ROLF NEVANLINNA
WHO CREATED THE BEAUTIFUL AND RICH FIELD OF VALUE DISTRIBUTION
THEORY OF MEROMORPHIC FUNCTIONS

CONTRIBUTORS TO THIS VOLUME


Chi-Tai Chuang
      The Department of Mathematics, Peking University, Beijing, China

Matts R. Essén
      The Department of Mathematics, University of Uppsala, Uppsala, Sweden

Albert Edrei
      The Department of Mathematics, Syracuse University, Syracuse, New York
      13210

Fred Gross
      The Department of Mathematics, University of Maryland at Baltimore County,
      Md. 21228 and Naval Research Laboratory, Washington, D.C.  20375

Yong-Xing Gu
      The Department of Mathematics, Nan-Chun Normal College, Sze-Chwan Province,
      China

Yu-Zan He
      Institute of Mathematics, Academia Sinica, Beijing, China

Simon Hellerstein
      The Department of Mathematics, University of Wisconsin, Madison, Wisconsin
      53706

Jun-Shung Hwang
      Institute of Mathematics, Academia Sinica, Taiwan

James K. Langley
      The Department of Mathematics, University of Illinois, Urbana, Illinois
      61801

Peter A. Lappan
      The Department of Mathematics, Michigan State University, Lansing,
      Michigan  48823

Lennox S.O. Liverpool
      The Department of Mathematics, Jos University, Jos, Nigeria

Seiki Mori
      The Department of Mathematics, Yamagata University, Yamagata, Japan

Kiyoshi Niino
      Faculty of Technology, Kanazawa University, Kanazawa 920, Japan

Charles F. Osgood
      Naval Research Laboratory, Washington, D.C.  20375

John Rossi
      The Department of Mathematics, Virginia Polytechnic Institute and State
      University, Blacksburg, Virginia  24061

Ranjan Roy
      The Department of Mathematics, Beloit College, Beloit, Wisconsin  53511

S.M. Shah
      The Department of Mathematics, University of Kentucky, Lexington,
      Kentucky  40506

Daniel F. Shea
>    The Department of Mathematics, University of Wisconsin, Madison,
>    Wisconsin  53706

Li-Chien Shen
>    The Department of Mathematics, The California Institute of Technology,
>    Pasadena, California  91125

Leonard M. Smiley
>    The Department of Mathematics, University of Alaska, Anchorage, Alaska
>    99506

Guo-Dong Song
>    The Department of Mathematics, Cornell University, Ithaca, New York
>    14853

Charles S. Stanton
>    The Department of Mathematics, University of North Carolina, Chapel Hill,
>    North Carolina  27514

Shlomo Strelitz
>    The Department of Mathematics, University of Haifa, Haifa, Israel

Nobushige Toda
>    The Department of Mathematics, Nagoya Institute of Technology, Nagoya,
>    Japan

Sakari Toppila
>    The Department of Mathematics, University of Helsinki, Helsinki 10,
>    Finland

Chia-Chi Tung
>    The Department of Mathematics, Astron. and Statistics, Mankato State
>    University, Mankato, Minnesota  56001

Jack Williamson
>    The Department of Mathematics, University of Hawaii, Honolulu, Hawaii
>    96822

Xiuzhi Xiao
>    The Department of Mathematics, Wuhan University, Wuhan, China

Nino Yanagihara
>    The Department of Mathematics, Chiba University, Chiba, Japan

Chung-Chun Yang
>    Naval Research Laboratory, Washington, DC  20375

Kenneth B. Yuguda
>    The Department of Mathematics, University of Jos, Jos, Nigeria

Contemporary Mathematics
Volume 25, 1983

## Generalizations of the Notion of Normal Families
## of Meromorphic Functions

Chi-tai Chuang

### Introduction

According to Montel [1], a family of meromorphic functions in a domain is said to be normal, if from every sequence of functions of the family one can extract a subsequence uniformly convergent in the interior of that domain. Next to this notion, Montel [1] introduced that of quasi-normality in which the extracted subsequence is allowed to have a set of points of nonuniform convergence, without point of accumulation in the interior of the domain. If we denote respectively by E and E' the set of points of nonuniform convergence and its derived set in the interior of the domain, then normality and quasi-normality correspond respectively to the conditions $E=\emptyset$, ... . In doing so, for each integer $m \geq 0$, we define a corresponding $Q_m$-normality corresponding to the condition $E^{(m)}=\emptyset$, such that in particular for $m=0$ and $m=1$, it coincides respectively with the notions of normality and quasi-normality introduced by Montel.

Many results on the theory of $Q_m$-normal families of meromorphic functions and its applications have been obtained.  However in the present paper we state and prove only some of them, with the purpose of justifying the introduction of such families.

### 1.  $Q_m$-normal families

Definition 1.  In what follows we denote the spherical distance (chordal) of two points $W_1$, $W_2$ of the extended complex plane $\hat{C}=C \cup (\infty)$ by $|W_1, W_2|$.  By a domain we mean an open connected set of points of the complex plane C.

Definition 2.  Let S: $f_n(z)$ (n=1,2,...) be a sequence of meromorphic functions in a domain D and $z_0$ a point of D.  We say that $z_0$ is a $C_0$-point of S, if there is a circle $\gamma: |z-z_0| < r$ interior to D such that S is spherically uniformly convergent in $\gamma$, namely to each positive number $\varepsilon$ corresponds a positive integer N such that for $n \geq N$, $m \geq N$, we have

$$|f_n(z), f_m(z)| < \varepsilon$$

in $\gamma$.  Next we say that $z_0$ is a $C_1$-point of S, if there exists a circle $|z-z_0|<r$ interior to D such that each point of the domain $0<|z-z_0|<r$ is a

$C_0$-point of S; $z_0$ is a $C_2$-point of S, if there exists a circle $|z-z_0|<r$ is a $C_1$-point of S.  In general, for an integer $m \geq 1$, after having defined the notion of $C_{m-1}$-point of S, we say that $z_0$ is a $C_m$-point of S, if there exists a circle $|z-z_0|<r$ interior to D such that each point of the domain $0<|z-z_0|<r$ is a $C_{m-1}$-point of S.  In this way, the notion of $C_m$-point of S is defined for each integer $m \geq 0$.

Lemma 1.  If, for an integer $m \geq 0$, $z_0$ is a $C_m$-point of S, then the following assertions hold:

$1^0$ We can find a circle $|z-z_0|<r$ interior to D such that each point of this circle is a $C_m$-point of S.

$2^0$ $z_0$ is a $C_{m+1}$-point of S.

$3^0$ $z_0$ is a $C_m$-point of every subsequence of S.

This lemma is easily proved by mathematical induction.

Definition 3.  $m \geq 0$ being an integer, we say that a sequence S of meromorphic functions in a domain D is a $C_m$-sequence in D, if each point of D is a $C_m$-point of S.

By known theorems [2], if S is a $C_0$-sequence in D, then S has a limit function defined in D, which is either a meromorphic function in D or the constant $\infty$; so in either case the limit function is spherically continuous in D.

Lemma 2.  If, for an integer $m \geq 0$, S is a $C_m$-sequence in D, then S is a $C_{m+1}$-sequence in D, and every subsequence of S is a $C_m$-sequence in D.

This lemma follows immediately from Lemma 1.

Definition 4.  Let S be a sequence of meromorphic functions in a domain D and $m \geq 0$ an integer.  A point $z_0$ of D is said to be a non-$C_m$-point of S, if $z_0$ is not a $C_m$-point of S.

Theorem 1.  Let S be a sequence of meromorphic functions in a domain D. If, for an integer $m \geq 0$, S is a $C_m$-sequence in D, then the set E of non-$C_0$-points of S in D is at most enumerable.

To prove this theorem we need the following lemma:

Lemma 3.  Let $\sigma$ be a set of points in a domain D.  If $\sigma$ has no point of accumulation in D, then $\sigma$ is at most enumerable.

We give here a proof of this lemma by a method which we have used in [3] and will be used later in other places.

First of all, let $z_j$ (j=1,2, ...) be a sequence of points of D, such that each point of D is a limiting point of the sequence $z_j$ (j=1,2, ...). We can get such a sequence in different ways, for example, we may take it to be the set of rational points[1] in D. It is well known that this set is enumerable.

Consider a point $z_j$ and let $R_j$ be the least upper bound of the set of positive numbers such that the circle $|z-z_j| < r$ is interior to D. Let $\Gamma_j$ be the circle $|z-z_j| < R_j/2$ if $R_j < +\infty$, and the circle $|z-z_j| < 1$, if $R_j = +\infty$. In this way, to each point $z_j$ is associated a circle $\Gamma_j$ with center $z_j$ and such that its closure $\bar{\Gamma}_j$ is interior to D.

Now consider a point z' of D and let $: |z-z'| < \rho$ $(0 < \rho < 1)$ be a circle interior to D. There is a point $z_j$ such that $|z_j - z'| < \rho'$, $\rho' = \rho/4$. Then the circle $|z-z_j| < 2\rho'$ is interior to $\gamma$. Consequently, by the definition of $R_j$, if $R_j < +\infty$, we have

$$2\rho' \leq R_j, \qquad \rho' \leq R_j/2$$

and the circle $|z-z_j| < \rho'$ is interior to the circle $\Gamma_j$; while, if $R_j = +\infty$, then since $\rho' < 1$, the circle $|z-z_j| < \rho'$ is also interior to $\Gamma_j$. But the point z' is a point of the circle $|z-z_j| < \rho'$, hence z' is a point of $\Gamma_j$. In particular, each point of the set $\sigma$ belongs to one of the circle $\Gamma_j$ (j=1,2,...). Now by hypothesis $\sigma$ has no point of accumulation in D and $\bar{\Gamma}_j \subset D$. It follows that each circle $\Gamma_j$ contains at most a finite number of points of $\sigma$. $\sigma$ is therefore at most enumerable.

Now let us return to the proof of theorem 1. For the case m=0, the set E is empty. For the case m-1, the set E has no point of accumulation in D, and hence at most enumerable, by Lemma 3.

To complete the proof, suppose that Theorem 1 is true for an integer $m \geq 1$ and consider a $C_{m+1}$-sequence S of meromorphic functions in a domain D. Let $\sigma$ be the set of non-$C_m$-points of S in D. Then $\sigma$ has no point of accumulation in D, and hence at most enumerable, by Lemma 3. On the other hand, S is a $C_m$-sequence in the domain $D_1 = D - \sigma$. Consequently, by assumption, the set

---

1) Points of the form a+bi where a and b are rational numbers.

$E_1$ of the non-$C_0$-points of S in D, is at most enumerable.  Evidently the set

$E \subset \sigma \cup E_1$, hence E is at most enumerable.  Theorem 1 is thus also true for the integer m+1.

Definition 5.  Let E be a set of points in a domain D.  Denote by E' the set of points of accumulation of E in the interior of D, by E" the set of points of accumulation of E' in the interior of D, in general, by $E^{(m)}$ the set of points of accumulation of $E^{(m-1)}$ in the interior of D.  If, for an integer $m \geq 0$, the sets $E^{(j)}$ (j=0,1,..., m; $E^{(0)}$=E) are all non-empty, we say that the set E has the property $W_m$ with respect to D.

Theorem 1 can be precised as follows:

Theorem 2.  Let S be a sequence of moeromorphic functions in a domain D, and $m \geq 0$ an integer.  In order that S is a $C_m$-sequence in D, it is necessary and sufficient that the set of non-$C_0$-points of S in D does not have the property $W_m$ with respect to D.

Proof:  The necessity of the condition can be proved as follows:  For m=0,1, the necessity is evident.  Now assume that the condition is necessary for an integer $m \geq 1$.  Consider a $C_{m+1}$-sequence S of meromorphic functions in a domain D, and suppose that the set E of non-$C_0$-points of S in D has the property $W_{m+1}$ with respect to D.  Let $z_0$ be a point of $E^{(m+1)}$ and $|z-z_0| < r$ a circle interior to D, such that in the domain d: $0 < |z-z_0| < r$ the sequence S is a $C_m$-sequence.  Then, since the sets $E^{(m)} \cap d$, $E^{(m-1)} \cap d$,..., $E \cap d$ are all non-empty, the set $E \cap d$ has the property $W_m$ with respect to d.  We are thus led to a contradiction with the assumption.

The sufficiency of the condition is also proved by mathematical induction as follows:  For m=0,1, the sufficiency is evident.  Assume that the condition is sufficient for an integer $m \geq 1$.  Suppose that a sequence S of meromorphic functions in a domain D is not a $C_{m+1}$-sequence in D.  Then there is a point $z_0$ of D, which is not a $C_{m+1}$-point of S.  Consider a circle $|z-z_0| < r$ interior to D.  Then S is not a $C_m$-sequence in the domain d: $0 < |z-z_0| < r$.  Then, by assumption, the set of non-$C_0$-points of S in d has the property $W_m$ with respect to d.  Consequently the set E of the non-$C_0$-points of S in D has the property $W_m$ with respect to D and the set $E^{(m)} \cap D$ is non empty.  Since r may be arbitrarily small, $z_0 \in E^{(m+1)}$.  Hence E has the property $W_{m+1}$ with respect to D.

Definition 6.  Let F be a family of meromorphic functions in a domain D and $m \geq 0$ an integer.  We say that the family F is $Q_m$-normal in D, if from every sequence of functions of the family F, we can extract a subsequence which is a $C_m$-sequence in D.

In particular, a $Q_0$-normal family is a normal family and a $Q_1$-normal family is a quasi-normal family, according to the definition given by Montel[1].

Lemma 4.  If, for an integer $m \geq 0$, a family F of meromorphic functions in a domain is $Q_m$-normal in D, then F is $Q_{m+1}$-normal in D.

This lemma is an immediate consequence of Lemma 2.

Definition 7.  Let F be a family of meromorphic functions in a domain D, $m \geq 0$, an integer and $z_0$ a point of D.  We say that the family F is $Q_m$-normal at the point $z_0$, if F is $Q_m$-normal in a circle $|z-z_0|<r$ interior to D.

Theorem 3.  Let F be a family of meromorphic functions in a domain D, and $m \geq 0$ an integer.  Then in order that the family F is $Q_m$-normal in D, it is necessary and sufficient that F is $Q_m$-normal at each point of D.

Proof:  It is easy to see that the condition is necessary.  We are going to prove the sufficiency of the condition by the method used in the proof of Lemma 3.

First let $z_j$ $(j=1,2,\ldots)$ be a sequence of points of D such that each point of D is a limiting point of the sequence of points $z_j$ $(j=1,2,\ldots)$.  Next consider a point $z_j$.  By hypothesis, there is a circle $|z-z_j|<r$ interior to D, in which the family F is $Q_m$-normal.  Let $R_j$ be the least upper bound of the numbers r having this property.  Let $\Gamma_j$ be the circle $|z-z_j|<R_j/2$, if $R_j < +\infty$, and the circle $|z-z_j|<1$, if $R_j=+\infty$.  $\Gamma_j$ is interior to D and in $\Gamma_j$ the family F is $Q_m$-normal.  Now let S: $f_n(z)$ $(n=1,2,\ldots)$ be a sequence of functions of the family F.  From S we can extract a subsequence

$$S_1: f_{\alpha_1}(z), f_{\alpha_2}(z), \ldots$$

which is a $C_m$-sequence in $\Gamma_1$.  From $S_1$ we can extract a subsequence

$$S_2: f_{\beta_1}(z), f_{\beta_2}(z), \ldots$$

which is a $C_m$-sequence in $\Gamma_2$.  From $S_2$ we can extract a subsequence

$$S_3: f_{\gamma_1}(z), f_{\gamma_2}(z), \ldots$$

which is a $C_m$-sequence in $\Gamma_3$.  In this way, we get successively a sequence of

sequence $S_p$ ($p=1,2,,..$) such that, for each $p \geq 1$, $S_p$ is a $C_m$-sequence in $\Gamma_p$ and

$S_{p+1}$ is a subsequence of $S_p$.  Consider the diagonal sequence

$$S': \quad f_{\alpha_1}(z), \; f_{\beta_2}(z), \; f_{\gamma_3}(z), \ldots$$

$S'$ is a subsequence of $S$ and, by Lemma 2, $S'$ is a $C_m$-sequence in each of the

circles $\Gamma_j$ ($j=1,2,\ldots$).

Consider a point $z'$ of D.  By hypothesis, there is a circle

$$\gamma: \left| z-z' \right| < \rho \qquad (0<\rho<1) \text{ interior to D, in which the}$$

family F is $Q_m$-normal.  Let $z_j$ be such that $\left| z_j - z' \right| < \rho'$, $\rho' = \rho/4$.  Then the

circle $\left| z-z_j \right| < 2\rho'$ is interior to $\gamma$ and hence F is $Q_m$-normal in the circle

$\left| z-z_j \right| < 2\rho'$.  Consequently, by the definition of $R_j$, if $R_j < +\infty$, we have

$$2\rho' \leq R_j, \quad \rho' \leq R_j/2$$

and the circle $\left| z-z_j \right| < \rho'$ is interior to $\Gamma_j$.  On the other hand, if $R_j = +\infty$, then,

since $\rho' < 1$, the circle $\left| z-z_j \right| < \rho'$ is also interior to $\Gamma_j$.  But $z'$ is a point of

the circle $\left| z-z_j \right| < \rho'$, hence $z'$ is a $C_m$-point of $S'$.  Since $z'$ is an arbitrary

point of D, $S'$ is a $C_m$-sequence in D.  Consequently the family F is $Q_m$-normal

in D.


$$2. \quad \text{Conditions of } Q_m\text{-normality}$$


Definition 8.  $m \geq 0$ being an integer, we say that a condition K is a

condition of $Q_m$-normality, when the following properties are true:

$1^0$  If a meromorphic function $f(z)$ in a domain D satisfies the condition

K in D, then $f(z)$ satisfies the condition K in every subdomain of D.[2]

$2^0$  If each function of a family F of meromorphic functions in a domain D

satisfies the condition K in D, then the family F is $Q_m$-normal in D.

Example 1.  Given a number $\delta$ ($0<\delta<1$), let $\kappa(\delta)$ be the corresponding

condition defined as follows:  We say that a meromorphic function $f(z)$ in a

domain D satisfies the condition $\kappa(\delta)$ in D, if there are three values

$a_j$ ($j=1,2,3$) such that

---

2)  It is tacitly assumed that for any domain D, there are functions
meromorphic and satisfying the condition K in D.

$$|a_j, a_{j'}| \geq \delta \quad (j,j'=1,2,3; \; j \neq j') \tag{1}$$

and that the function $f(z)$ does not take the values $a_j$ $(j=1,2,3)$ in D. By a known theorem [2], $\kappa(\delta)$ is a condition of $Q_0$-normality.

Definition 9. $m \geq 0$ being an integer, let K be a condition of $Q_m$-normality and $\varepsilon$ a positive number. Let $f(z)$ be a meromorphic function in a domain D. A circle $\gamma: |z-z_0| < r$ is called a $(K,\varepsilon)$-circle of the function $f(z)$ in D, if the following properties are true:

$1^0$ $\bar{\gamma}: |z-z_0| < r$ is interior to D.

$2^0$ $r < \varepsilon$.

$3^0$ The function $f(z)$ does not satisfy the condition K in $\gamma$. Several $(K,\varepsilon)$-circles $\gamma_j$ $(j=1,2,\ldots,p)$ are said to be distinct, if their closures $\bar{\gamma}_j$ $(j=1,2,\ldots,p)$ have no common points each other.

Example 2. $\kappa(\delta)$ being the condition of $Q_0$-normality defined in example 1, a $(\kappa(\delta),\varepsilon)$-circle $\gamma$ of a meromorphic function $f(z)$ in a domain D, is such that among any three values $a_j$ $(j=1,2,3)$ satisfying the condition (1), there is at least one $a_j$ which the function $f(z)$ takes in $\gamma$. Consequently the function $f(z)$ takes in $\gamma$ every value W except at most those values belonging to two spherical circles $|W, W_j| < \delta (j=1,2)$. The circle $\gamma$ is therefore a filling circle of the function $f(z)$.

Theorem 4. $m \geq 0$ being an integer, let K be a condition of $Q_m$-normality and $\varepsilon$ a positive number. Let F be a family of meromorphic functions in a domain D. If each function of the family F has no $(K,\varepsilon)$-circle in D, then the family F is $Q_m$-normal in D.

Proof: Consider a point $z_0$ of D and let $\gamma: |z-z_0| < r$ be a circle having the first two properties in Definition 9. Then, by hypothesis, each function of the family F satisfies the condition K in $\gamma$. Hence the family F is $Q_m$-normal in $\gamma$. Since $z_0$ is arbitrary, by Theorem 3, F is $Q_m$-normal in D.

Definition 10. $m \geq 1$ being an integer, a family F of meromorphic functions in a domain D is said to be $Q_m$-normal of order $\nu$ $(\nu \geq 0$ integer) at most in D, if from every sequence of functions of the family F, we can extract a subsequence which is a $C_m$-sequence in D and has at most $\nu$ non-$C_m$-points in D.

Definition 11.  $m \geq 1$ and $\nu \geq 0$ being integers, we say that a condition K is a condition of $Q_m$-normality of order $\nu$ at most, when the following properties are true:

$1^0$  The first property in definition 8.

$2^0$  If each function of a family F of meromorphic functions in a domain D satisfies the condition K in D, then the family F is $Q_m$-normal of order $\nu$ at most in D.

Theorem 5.  $m \geq 0$ being an integer, consider a condition K of $Q_m$-normality, a positive number $\varepsilon$ and an integer $\nu \geq 0$.  Let K' be the corresponding condition defined as follows:  We say that a meromorphic function $f(z)$ in a domain D satisfies the condition K' in D, if the function $f(z)$ has at most $\nu$ distinct $(K,\varepsilon)$-circles in D.  Then the condition K' is a condition of $Q_{m+1}$-normality of order $\nu$ at most.

Proof:  Evidently it is sufficient to show that if a family F of meromorphic functions in a domain D is such that each function of the family F satisfies the condition K' in D, then the family F is $Q_{m+1}$-normal of order $\nu$ at most in D.

This is immediate in the case $\nu=0$, because then each function of the family F has no $(K,\varepsilon)$-circle in D, hence, by Theorem 4, the family F is $Q_m$-normal in D.

Now consider the case $\nu \geq 1$.  Let $f_n(z)$ $(n=1,2,\ldots)$ be a sequence of functions of the family F and distinguish two cases:

$1^0$  For each point $z_0$ of D, there is a circle $\gamma: |z-z_0| < r$ interior to D and a positive integer N such that, for $n \geq N$, the function $f_n(z)$ satisfies the condition K in $\gamma$.  Then since K is a condition of $Q_m$-normality, the family $\{f_n(z), n=1,2,\ldots\}$ is $Q_m$-normal in $\gamma$.  Hence, by Theorem 3, this family is $Q_m$-normal in D and we can extract a subsequence $f_{n_k}(z)$ $(k=1,2,\ldots)$ which is a $C_m$-sequence in D.

$2^0$  There exists a point $z_0$ of D such that whatever be the circle $\gamma: |z-z_0| < r$ interior to D and the positive integer N, we can always find an integer $n \geq N$ such that the function $f_n(z)$ does not satisfy the condition K in $\gamma$.  For the sake of simplicity, such a point $z_0$ will be called an exceptional point of the sequence of functions $f_n(z)$ $(n=1,2,\ldots)$.  We are going to show that we can extract from the sequence $f_n(z)$ $(n=1,2,\ldots)$, a subsequence $f_{\alpha_i}(z)$ $(k=1,2,\ldots)$ having the following property $(\pi)$ with respect to the exceptional point $z_0$: For any circle $\gamma: |z-z_0|<r$ interior to D, the function $f_{\alpha_i}(z)$ does not satisfy the condition K in $\gamma$, provided that i is sufficiently large. In fact, take a sequence of positive numbers $r_i$ $(i=1,2,\ldots)$ tending to zero, such that the circles $\gamma_i: |z-z_0|<r_i$ $(i=1,2,\ldots)$ are all interior to D. Then we can find an increasing sequence of positive integers $\alpha_i$ $(i=1,2,\ldots)$ such that for each i the function $f_{\alpha_i}(z)$ does not satisfy the condition K in $\gamma_i$. It is easy to see that the sequence $f_{\alpha_i}(z)$ $(i=1,2,\ldots)$ has the required property $(\pi)$.

The point $z_0$ being an exceptional point of the sequence of functions $f_{\alpha_i}(z)$ $(i=1,2,\ldots)$, suppose that it has another exceptional point $z_0' \neq z_0$ in D. Then from the sequence of functions $f_{\alpha_i}(z)$ $(i=1,2,\ldots)$ we can extract a subsequence $f_{\beta_j}(z)$ $(j=1,2,\ldots)$ having the property $(\pi)$ with respect to the point $z_0'$. Evidently the sequence of functions $f_{\beta_j}(z)$ $(j=1,2,\ldots)$ also has the property $(\pi)$ with respect to the point $z_0$.

The points $z_0$ and $z_0'$ being exceptional points of the sequence of functions $f_{\beta_j}(z)$ $(j=1,2,\ldots)$, suppose that it has another exceptional point $z_0'' \neq z_0, z_0'$ in D. Then from the sequence of functions $f_{\beta_j}(z)$ $(j=1,2,\ldots)$ we can extract a subseuqence $f_{\gamma_k}(z)$ $(k=1,2,\ldots)$ having the property $(\pi)$ with respect to the point $z_0''$. The sequence of functions $f_{\gamma_k}(z)$ $(k=1,2,\ldots)$ also has the property $(\pi)$ with respect to $z_0$ and $z_0'$.

If this operation can be repeated S times, then we get a subsequence $f_{\lambda_1}(z)$ $(1=1,2,\ldots)$ of the sequence of functions $f_n(z)$ $(n=1,2,\ldots)$ and S distinct points $z_0, z_0', \ldots, z_0^{(s-1)}$ of D, such that the sequence of functions $f_{\lambda_1}(z)$ $(1=1,2,\ldots)$ has the property $(\pi)$ with respect to each of the points $z_0, z_0', \ldots, z_0^{(s-1)}$. We are going to show that $S \leq \nu$, where $\nu$ is the integer occurring in Theorem 5. In fact, $\varepsilon$ being the positive number occurring in the same theorem, let us take a positive number $r < \varepsilon$ such that the circles

$\bar{\gamma}_j : |z - z_0^{(j)}| \leq r$ $(j = 0, 1, \ldots, s-1; z_0^{(0)} = z_0)$ are interior to D and without common

point each other.  Then we can get a positive integer i such that in each of

the circles $\gamma_j$ $(j = 0, 1, \ldots, s-1)$, the function $f_{\lambda_i}(z)$ does not satisfy the

condition K.  But by hypothesis, $f_{\lambda_i}(z)$ satisfies the condition K' in D, so

we must have $s \leq \nu$ .

Thus for a certain positive integer $S \leq \nu$, the subsequence $f_{\lambda_i}(z)$

$(i = 1, 2, \ldots)$ obtained in repeating the operation S times, has no other

exceptional point distinct from the S points $z_0, z_0', \ldots, z_0^{(s-1)}$ .  Then in

the domain $G = D - (z_0, z_0', \ldots, z_0^{(s-1)})$ the sequence of functions $f_{\lambda_i}(z)$

$(i = 1, 2, \ldots)$ satisfies the condition of the first case.  Hence we can extract

from this sequence of functions, a subsequence $f_{n_k}(z)$ $(k = 1, 2, \ldots)$ which is a

$C_m$-sequence in G.  Consequently the sequence functions $f_{n_k}(z)$ $(k = 1, 2, \ldots)$ is a

$C_{m+1}$-sequence in D, of which only the $S \leq \nu$ points $z_0, z_0', \ldots, z_0^{(s-1)}$ may be

non-$C_m$-points.

The conditions K' defined in Theorem 5 depends on m, K, $\varepsilon$ and $\nu$.  We
express this dependence in writing

$$K' = \Omega(m, K, \varepsilon, \nu). \tag{2}$$

f(z) being a meromorphic function in a domain D, then, according to Definition
9, a $(K', \varepsilon')$-circle of the function f(z) in D is a circle $\gamma : |z - z_0| \leq r$ with

the following properties:

$1^0$ $\bar{\gamma} : |z - z_0| \leq r$ is interior to D.

$2^0$ $r < \varepsilon$.

$3^0$ The function f(z) has at least $\nu + 1$ distinct $(K, \varepsilon)$-circles in $\gamma$.

Now consider a condition $K_0$ of $Q_0$-normality and a sequence $(\varepsilon_m, \nu_m)$

$(m = 0, 1, 2, \ldots)$, where for each m, $\varepsilon_m$ is a positive number and $\nu_m$ a non negative

integer.  Then, by Theorem 5,

$$K_1 = \Omega(0, K_0, \varepsilon_0, \nu_0)$$

is a condition of $Q_1$-normality of order $\nu_0$ at most;

$$K_2 = \Omega(1, K_1, \varepsilon_1, \nu_1)$$

is a condition of $Q_2$-normality of order $\nu_1$ at most; in general,

$$K_m = \Omega(m-1, K_{m-1}, \varepsilon_{m-1}, \nu_{m-1}) \quad (m=1,2,\ldots) \tag{3}$$

is a condition of $Q_m$-normality of order $\nu_{m-1}$ at most.  We say that the sequence of conditions $K_m$ $(m=0,1,2,\ldots)$ is generated by the condition $K_0$ and the sequence $(\varepsilon_m, \nu_m)$ $(m=0,1,2,\ldots)$.

Example 3.  The condition $\kappa(\delta)$ of $Q_0$-normality defined in example 1 and a sequence $(\varepsilon_m, \nu_m)$ $(m=0,1,2,\ldots)$ generate a sequence of conditions $\kappa_m$ $(m=0,1,2,\ldots; \ \kappa_0 = \kappa(\delta))$.

## 3.  An application

Just as the theory of normal families, the theory of $Q_m$-normal families has various applications.  However in the present paper, as an application, we prove only the following theorem which is a generalization of a theorem of Montel [1].

Theorem 6.  Let S: $f_n(z)$ $(n=1,2,\ldots)$ be a sequence of holomorphic functions in a domain D and E a set of points of D.  Suppose that, for an integer $m \geq 1$, S belongs to a $Q_m$-normal family of holomorphic functions in D and E has the property $W_m$ with respect to D.  Moreover suppose that, at each point of E, S converges to a finite limit.  Then S converges locally uniformly to a holomorphic function in D.

For the proof of this theorem, we need the following lemma:

Lemma 5.  Under the conditions of Theorem 6, we can extract from S a subsequence S' which converges locally uniformly to a holomorphic function in D.

Proof:  Consider first the case $m=1$.  In this case, S belongs to a $Q_1$-normal family of holomorphic functions in D.  Hence, by defnition, we can extract from S a subsequence S' which is a $C_1$-sequence in D.  Denote by $\sigma$ the set of non-$C_0$-points of S' in D.  Then S' is a $C_0$-sequence in the domain $G=D-\sigma$.  Let $z_0$ be a point of E' and $\gamma: |z-z_0| < r$ a circle such that its closure $\bar{\gamma} \subset D$.  Since $\sigma$ has no point of accumulation in D, $\gamma$ contains at most a finite number of points of $\sigma$, but an infinite number of points of E, hence there is a point $z_1 \varepsilon$ E$\cap$G.  By hypothesis, at the point $z_1$, the limit of S' is finite; Since S' is a $C_0$-sequence of holomorphic functions in G, S' is locally uniformly convergent in G [3].  Next consider a point $\zeta_0 \varepsilon \sigma$.  Let

$|z-\zeta_0|<\rho$ be a circle interior to D, such that each point of the domain $0<|z-\zeta_0|<\rho$ is a $C_0$-point of S'.  Then the domain $0<|z-\zeta_0|<\rho$ belongs to G, hence S' is uniformly convergent on the circle $|z-\zeta_0|=\rho/2$ .  It follows, by the principle of maximum modulus, that S' is uniformly convergent in the circle $|z-\zeta_0|<\rho/2$.  S' is therefore locally uniformly convergent in D.  This proves Lemma 5 in the case m=1.

To complete the proof of Lemma 5, suppose that it is true for an integer m$\geq$1, and consider the case corresponding to the integer m+1.  In this case, we can extract from the sequence S a subsequence S' which is a $C_{m+1}^-$ sequence in D.  Let $\sigma$ be the set of non-$C_m$-points of S in D.  Then S' is a $C_m$-sequence in the domain $D_1$=D-$\sigma$ and, a fortiori, S' considered as a family of functions is $Q_m$-normal in $D_1$.  By considering a point $Z_0 \varepsilon E^{(m+1)}$, we see as above that there is a point $Z_1 \varepsilon E^{(m)} \cap D_1$.  Then it is easy to see that the set $E_1=E \cap D_1$ has the property $W_m$ with respect to $D_1$.  Moreover S' converges to a finite limit at each point of $E_1$.  Hence, by assumption, we can extract from S', a subsequence S" which is locally uniformly convergent in $D_1$.  Then we see, as above, that S" is locally uniformly convergent in D.  The proof of Lemma 5 is now complete.

Now let us prove Theorem 6.  First of all, by Lemma 5, we can extract from S a subsequence S' which converges locally uniformly to a holomorphic function F(z) in D.  We are going to show that S converges locally uniformly to the function F(z) in D.  In fact, consider a circle $\gamma: |z-z_0|<r$ such that its closure $\bar{\gamma} \subset$ D.  Suppose that S does not converge uniformly to the function F(z) in $\bar{\gamma}$.  Then we can find a positive number $\varepsilon_0$ and an increasing sequence of positive integers $n_k$ (k=1,2,...) such that, for each $\kappa \geq 1$, we have

$$\max_{\bar{\gamma}} |g_k(z)-F(z)| \geq \varepsilon_0, \tag{4}$$

where

$$g_k(z) = f_{n_k}(z) \qquad (k=1,2,\ldots).$$

Consider the sequence of functions $S_1$: $g_k(z)$ $(k=1,2,\ldots)$. Since $S_1$ satisfies

the same conditions in Theorem 6, hence, by Lemma 5, we can also extract

from $S_1$ a subsequence $S_1'$: $g_{k_\lambda}(z)$ $(\lambda=1,2,\ldots)$ which converges locally

uniformly to a holomorphic function $F_1(z)$ in D. There is then a positive

integer $\lambda_0$ such that, for $\lambda \geq \lambda_0$, the inequality

$$\left| g_{k_\lambda}(z) - F_1(z) \right| < \varepsilon_0 \tag{5}$$

holds in $\overline{\gamma}$. But on the set E, we have

$$F(z) = F_1(z) \tag{6}$$

and, by hypothesis, the set E' is non-empty; it follows that the identity (6)

holds in D. The inequalities (4) and (5) being incompatible, so we get a

contradiction.

## 4.  An existence theorem

It is natural to ask the following question:  Given arbitrarily an

integer $m \geq 1$, can we always find a family of meromorphic functions in a domain

D, which is $Q_m$-normal in D, without being $Q_{m-1}$-normal in D? We are going to

give an affirmative answer to this question. First of all, we point out that,

for this purpose, it is sufficient to prove the following theorem:

Theorem 7.  Given any integer $m \geq 1$, we can find a sequence S of

meromorphic functions in a domain D, such that S is a $C_m$-sequence in D, with-

out being a $C_{m-1}$-sequence in D, and that the same is true for every subse-

quence of S.

In fact, such a sequence S of meromorphic functions in a domain,

considered as a family of meromorphic functions in the same domain, evidently

satisfies the required conditions in the above question.

To prove Theorem 7, consider first the case m=1. In this case, the

construction of S is known and easy. In fact, the sequence of functions

$$f_n(z) = nz \qquad (n=1,2,\ldots)$$

are holomorphic in C and has only one non-$C_0$-point, namely the point $z=0$.

Hence this is a $C_1$-sequence in C, without being a $C_0$-sequence in C. This is also true for each of its subsequences.

In order to prove Theorem 7 for the case $m \geq 2$, we need the following two lemmas.

Lemma 6.   Let E be a set of points of C satisfying the following conditions:

$1^0$   E is infinite and enumerable.

$2^0$   $E \cap E'$ is empty.

Then we can construct a sequence S of holomorphic functions in C such that the set $E \cup E'$ is precisely the set of non-$C_0$-points of S in C, and that the same is true for every subsequence of S.

Proof:   Since E is enumerable, it can be represented by a sequence of points $a_n$ $(n=1,2,\ldots)$.   By condition $2^0$, to each point $a_n$ corresponds a number $0<\delta_n<1$ such that the domain $0<|z-a_n|<2\delta_n$ contains no point of E, consequently

$$|a_m-a_n|>2\delta_n \quad \text{for } m \neq n \tag{7}$$

We may suppose that $\delta_{n+1}<\delta_n$ $(n=1,2,\ldots)$.   Next take two sequences of numbers $\rho_n$ $(n=1,2,\ldots)$ and $d_n$ $(n=1,2,\ldots)$ such that

$$\rho_n>0 \ (n=1,2,\ldots), \ \lim_{n \to +\infty} \rho_n=+\infty \tag{8}$$

$$0<d_n<1 \ (n=1,2,\ldots), \ \lim_{n \to +\infty} d_n=0 \tag{9}$$

and construct a sequence of functions S: $f_n(z)$ $(n=1,2,\ldots)$ as follows:

$$f_1(z)= \frac{P_1}{d_1\delta_1}(z-a_1)$$

$$f_2(z)= \frac{P_2}{d_2^2\delta_2^2}(z-a_1)\ (z-a_2)$$

$$\cdot \ \cdot \ \cdot \ \cdot \ \cdot \ \cdot \ \cdot \ \cdot \ \cdot \ \cdot$$

$$f_n(z) = \frac{P_n}{d_n^n\delta_n^n}(z-a_1)\ (z-a_2)\ldots(z-a_n)$$

$$\cdot \ \cdot \ \cdot \ \cdot \ \cdot \ \cdot \ \cdot \ \cdot \ \cdot$$

Consider the point $a_1$ and a circle $C_1$: $|z-a_1|<\delta$ with $\delta<\delta_1$.   By (7), if $m \neq 1$, then in $C_1$ we have

$$|z-a_m| \geq |a_1-am| - |z-a_1| > 2\delta_1 - \delta > 2\delta_1 - \delta_1 = \delta_1.$$

It follows that in $C_1$ we have

$$|f_n(z)| \geq \frac{P_n}{\delta_1^n} |(z-a_1)(z-a_2)\ldots(z-a_n)| \geq \frac{P_n}{\delta_1}| z-a_1| \geq P_n| z-a_1|.$$

By (8), this inequality shows that $a_1$ is a non-$C_0$-point of S.  In general,

consider a point $a_n$ and a circle $C_n: |z-a_n| < \delta$ with $\delta < \delta_n$.  By (7), if $m \neq n$, then

in $C_n$ we have

$$|z-a_m| \geq |a_n-a_m| - |z-a_n| > 2\delta_n - \delta > 2\delta_n - \delta_n = \delta_n.$$

It follows that, for any integer $k \geq 0$, we have in $C_n$,

$$|f_{n+k}(z)| \geq P_{n+k}|z-a_n|, \tag{10}$$

and hence $a_n$ is a non-$C_0$-point of S.

Thus we have shown that each point of E is a non-$C_0$-point of S.  Of

course each point of E' (if E' is non-empty) is also a non-$C_0$-point of S.

Now consider a point $Z_0 \in \overline{E \cup E'}$.  Then we can find a circle $|z-z_0| < r$

which contains no point of E.  Consider the circle $C: |z-z_0| < r/2$.  For $m \geq 1$, we

have in C,

$$|z-a_m| > |z_0-a_m| - |z-z_0| > r - \frac{r}{2} = \frac{r}{2}.$$

Let N be a positive integer such that $d_n < r/2$ for $n \geq N$.  Then, for $n \geq N$, we have

in C,

$$|f_n(z)| > P_n$$

Hence $z_0$ is a $C_0$-point of S.

It is then proved that the set $E \cup E'$ is precisely the set of non-$C_0$-points of

S.  This is also true for any subsequence $f_{n_j}(z)$ (j=1,2,...) of S, because,

from (10), we have in $C_n$,

$$|f_{n_j}(z)| \geq P_{n_j}| z-a_n|$$

provided that j is sufficiently large.

Lemma 7.  Let $p \geq 1$ be an integer and $a < b$ be two real numbers.  Then we can find a set of points $E$ of the closed interval $[a,b]$ satisfying the following conditions:

$1^0$  $E$ is infinite and enumerable.

$2^0$  $E \cap E'$ is empty.

$3^0$  The set $E \cup E'$ has the property $W_p$ with respect to $C$ and $(E \cup E')^{(p+1)}$ is empty.

Proof:  When $p=1$, it is sufficient to take the sequence of points

$$E: \quad x_n = a + \frac{1}{n}(b-a) \qquad (n=1,2,\ldots) \tag{11}$$

Evidently $E$ satisfies the conditions $1^0$ and $2^0$.  Moreover, we have

$$E' = (a), \quad (E \cup E')' = (E \cup (a))' = a,$$

hence the condition $3^0$ is also satisfied.

Now suppose that Lemma 7 is proved for an integer $p \geq 1$.  Consider the integer $p+1$ and a closed interval $[a,b]$.  Let $x_n$ $(n=1,2,\ldots)$ be defined by (11) and consider the sequence of closed intervals:  $I_j : [x_{2j}, x_{2j-1}]$ $(j=1,2,\ldots)$.  By hypothesis, to each $j$ corresponds a set of points $E_j$ of the closed interval $I_j$ satisfying the conditions $1^0$, $2^0$, $3^0$ for the integer $p$.  Consider the set of points $E = \bigcup_{j=1}^{\infty} E_j$ of the closed interval $[a,b]$.

It is easy to see that $E$ satisfies the conditions $1^0$ and $2^0$.  On the other hand, it is also easy to see that

$$E' = \left( \bigcup_{j=1}^{\infty} E_j' \right) \cup (a)$$

which implies that  .

$$E \cup E' = \left( \bigcup_{j=1}^{\infty} H_j \right) \cup (a),$$

where $H_j = E_j \cup E_j'$.  Since $H_j \subset I_j$, we see that

$$(E \cup E')' = \left( \bigcup_{j=1}^{\infty} H_j \right)' \cup (a) = \left( \bigcup_{j=1}^{\infty} H_j' \right) \cup (a).$$

Then we have successively

$$(E \cup E')'' = \left( \bigcup_{j=1}^{\infty} H_j'' \right) \cup (a)$$

. . . . . . . . . .

$$(E \cup E')^{(p)} = (\bigcup_{j=1}^{\infty} H_j^{(p)}) \cup (a).$$

Since $H_j^{(p+1)}$ is empty $(j=1,2,\ldots)$, we see that

$$(E \cup E')^{(p+1)} = (a).$$

Consequently the set $E \cup E'$ has the property $W_{p+1}$ with respect to C and

$(E \cup E')^{(p+2)}$ is empty.  Therefore Lemma 7 is also true for the integer p+1.

Now let us return to the proof of Theorem 7 for the case $m \geq 2$.  Given an integer $m \geq 2$, consider the integer $p=m-1 \geq 1$.  By Lemma 7, we can find a set of points E of C, satisfying the conditions $1^0$, $2^0$, $3^0$ in Lemma 7.  From the first conditions, by Lemma 6, we can construct a sequence S of holomorphic functions in C wuch that the set $E \cup E'$ is precisely the set of non-$C_0$-points of S in C and that the same is true for every subsequence of S.  Since the set $E \cup E'$ satisfies the condition $3^0$ in Lemma 7, then by Theorem 2, S is a $C_{p+1}$-sequence in C, without being a $C_p$-sequence in C, and the same is true for every subsequence of S.  Since $p+1=m$, $p=m-1$, S satisfies the conditions required by Theorem 7, with D = C.

## 5.  Further extensions

Immediately after the integers, 0, 1, 2, . . . is the transfinite number $\omega$.  Let S be a sequence of meromorphic functions in a domain D.  We say that a point $z_0$ of D is a $C_\omega$-point of S, if there exists an integer $m \geq 0$, such that $z_0$ is a $C_m$-point of S.  If each point of D is a $C_\omega$-point of S, then we say that S is a $C_\omega$-sequence in D.  A family F of meromorphic functions in a domain D is said to be $Q_\omega$-normal in D, if from every sequence of functions of F, we can extract a subsequence which is a $C_\omega$-sequence in D.  Many results concerning $Q_m$-normal families ($m \geq 0$ integer) can be extended to $Q_\omega$-normal families.

We can go still further.  S being a sequence of meromorphic functions in a domain D, a point $z_0$ of D is said to be a $C_{\omega+1}$-point of S, if there is a circle $|z-z_0| < r$ interior to D such that each point of the domain $0| < z-z_0| < r$ is a $C_\omega$-point of S.  Then we define the terms $C_{\omega+1}$-sequence and $Q_{\omega+1}$-normal family by the same procedure used above, and so on.

C. CHUANG

## References


1. Montel, P., Lecons sur les familles normales de fonctions analytiques et leurs applications, Paris, 1927.

2. Valiron, G., Familles normales et quasi-normales de fonctions meromorphes, Memorial des Sc. math., fasc. 38.

3. Chuang, C.T., Singular directions of meromorphic functions (in Chinese), Peking, 1982.



Department of Mathematics
Peking University
Beijing, China

Contemporary Mathematics
Volume **25**, 1983

# ZEROS OF SECTIONS OF POWER SERIES REPRESENTING ENTIRE
# FUNCTIONS ADMISSIBLE IN THE SENSE OF HAYMAN

Albert Edrei

**Introduction.** Let

$$f(z) = \sum_{j=0}^{\infty} a_j z^j \tag{1}$$

be entire and transcendental.

This note is devoted to one aspect of the following

**Question.** **Where are the zeros of the partial sums**

$$s_m(z) = \sum_{j=0}^{m} a_j z^j, \tag{2}$$

**for large values of** m.

A first approximation of the position of these zeros was obtained (for functions of positive or infinite order) by Carlson [2], [3] and Rosenbloom [9], [10].

Special choices of f(z) lead to results considerably more informative than the general assertions of Carlson and Rosenbloom.

In particular, the masterful study by Szegö [11] of

$$f(z) = e^z$$

yields statements of great elegance and complete precision.

Results as precise as those of Szegö have recently been obtained by Edrei, Saff and Varga [6] for the Mittag-Leffler functions of all orders $\lambda > 1$:

$$E_{1/\lambda}(z) = \sum_{j=0}^{\infty} \frac{z^j}{\Gamma(1+\frac{j}{\lambda})} \qquad (1 < \lambda < +\infty).$$

Other classes have been studied by the present author [4]. One of the classes which he considers contains the function $1/\Gamma(z)$.

Regarding functions of infinite order, nothing seems to be known beyond the assertions of Carlson and Rosenbloom.

The objective of this note is to outline a method which permits the complete and precise treatment of

$$e_1(z) = \exp(z) = \exp(e_0(z)) \qquad (e_0(z) \equiv z),$$

and, more generally, of all the iterates

$$e_k(z) = \exp(e_{k-1}(z)) \qquad (k=1,2,3,\ldots). \tag{3}$$

The method in question rests on the remarkable properties, discovered by Hayman [7], of a class of functions which he calls "admissible."

Among the many properties of admissible functions, proved by Hayman [7; p. 69], we state some of the simplest:

If $f(z)$ and $g(z)$ are admissible functions, and $P(z)$ is a real polynomial, then the functions

$$f(z)g(z), \quad \exp(f(z)), \quad f(z) + P(z),$$

are also admissible. Admissibility also holds for

$$P(f(z)), \quad f(z)P(z),$$

provided the leading coefficient of $P(z)$ is positive.

The above properties will enable us to apply Theorem 1 below to classes of functions considerably more general than the class of iterated exponentials.

Definition of Hayman's admissible functions. Let $f(z)$ be a real, entire transcendental function such that

$$f(r) > 0 \qquad (r > r_0 > 0). \tag{4}$$

Put

$$a(r) = r\,\frac{f'(r)}{f(r)}, \quad b(r) = ra'(r), \tag{5}$$

and assume that

$$b(r) \to +\infty \qquad (r \to +\infty). \tag{6}$$

In addition assume that there exists some function $\delta(r)$ defined for $r > r_0$, such that

$$0 < \delta(r) < \pi, \tag{7}$$

and such that

$$f(re^{i\theta}) \sim f(r)\,\exp\!\left(i\theta a(r) - \tfrac{1}{2}\theta^2 b(r)\right) \qquad (r \to +\infty), \tag{8}$$

uniformly for

$$|\theta| \leq \delta(r), \tag{9}$$

while, uniformly for

$$\delta(r) \leq |\Theta| \leq \pi, \tag{10}$$

we have

$$f(re^{i\Theta}) = o(f(r))\{b(r)\}^{-\frac{1}{2}} \qquad (r \to +\infty). \tag{11}$$

An entire function having all the above properties is said to be admissible (in the sense of Hayman).

As an immediate consequence of (5) and (6) we note that

I.  The function $a(r)$ is a positive, continuous, strictly increasing, unbounded function of $r > r_0$.

The asymptotic behavior described by (8) and (11) clearly implies

$$\max_{|z|=r} |f(z)| = M(r) \sim f(r) \qquad (r \to +\infty). \tag{12}$$

Hayman states his fundamental result as follows:

II.  Let $f(z)$, given by its expansion (1), be admissible and define $a_j = 0$ for $j < 0$.

Then, as $r \to +\infty$, we have, uniformly for all integers $j$,

$$a_j r^j = f(r)\{2\pi b(r)\}^{-\frac{1}{2}} \left\{ \exp\left(-\frac{(a(r)-j)^2}{2b(r)}\right) + o(1) \right\}. \tag{13}$$

By property I, it is always possible, given $j > j_0 > 0$ to define a unique $r_j > 0$

$$a(r_j) = j \tag{14}$$

and hence (13) implies

III.  All but a finite number of coefficients $a_j$ $(j > 0)$ are positive.

Adopting the above definitions and notations we prove

Theorem 1.  Let $f(z)$ be an entire function, given by its expansion (1), and admissible in the sense of Hayman.

Assume in addition that, for some $\alpha$ $(0 < \alpha < \frac{1}{4})$, $f(z)$ has no zeros in any of the disks

$$S(R) = \{z: |z-R| \leq R\{a(R)\}^{-\alpha} \qquad (R > R_0 > 0), \tag{15}$$

and that

$$a(x) > (\log x)^{\beta} \qquad (\beta > \frac{3}{1-4\alpha}, \ x > x_0). \tag{16}$$

Assume also that

$$b'(x) \geq 0 \qquad (x > x_0 > 0). \tag{17}$$

Then, if R avoids some exceptional set E, of finite logarithmic measure, and if

$$m = m(R) = [a(R)], \tag{18}$$

we have, uniformly on every compact subset of the $\zeta$-plane,

$$\{1 + \zeta(\frac{2}{b(R)})^{1/2}\}^{-m} \{f(R)\}^{-1} s_m(R\{1+\zeta(\frac{2}{b(R)})^{1/2}\}) \rightarrow \tag{19}$$

$$\frac{1}{2} e^{\zeta^2} (1 - \frac{2}{\sqrt{\pi}} \int_0^{\zeta} e^{-\sigma^2} d\sigma) \qquad (R \rightarrow +\infty, \ R \notin E, \ m=m(R)).$$

Remark 1. If the growth of $a(r)$ is regular enough to always imply

$$a(r + \frac{r}{\{a(r)\}^{\alpha}}) < 2a(r) \qquad (r > r_0), \tag{20}$$

the introduction of the exceptional set E becomes unnecessary.

Remark 2. If (20) holds with r=R, then, since $b(x)$ is positive and non-decreasing (by (17) and (6)), we find

$$\frac{1}{2} b(R)\{a(R)\}^{-\alpha} \leq \int_R^{R+R\{a(R)\}^{-\alpha}} \frac{b(t)}{t} dt = a(R+R\{a(R)\}^{-\alpha}) - a(R) < a(R)$$

and hence

$$b(R) \leq 2\{a(R)\}^{1+\alpha} \qquad (R > R_0, \ R \notin E). \tag{21}$$

An inequality in the opposite sence is immediate:

$$a(r) - a(r_0) = \int_{r_0}^{r} \frac{b(t)}{t} dt < b(r)(\log r - \log r_0),$$

and hence

$$a(r) < 2b(r) \log r \qquad (r > r_0 > 0). \tag{22}$$

Remark 3. A condition such as (16) precludes the appearance of certain functions of order zero. If the condition were completely omitted, we could not rule out functions such as

$$\sum_{n=0}^{\infty} a^{-n^2} z^n \qquad (a \geq 2),$$

Theorem 1 cannot hold for such functions because, as is well-known [8; p. 69, ex 176] the zeros of all the partial sums of the above series are real and negative.

Remark 4.  Let g(z) be _admissible in the sense of Hyaman_.  Nothing is assumed regarding the zeros of g(z), nor do we assume that the functions a(x) and b(x) associated with g(z) satisfy the conditions (16) and (17).

Then

$$f(z) = \exp(g(z)) \tag{23}$$

satisfies all the conditions of Theorem 1.

This remark clearly implies that Theorem 1 may be applied to all the iterated exponentials $e_k(z)$  $(k \geq 2)$.  For sake of completeness we sketch, at the end of the paper [§7], a brief proof of this remark.

A striking feature of Theorem 1 is the universal character of the limit function in (19).  This function, has zeros which coincide with those of the well-known complementary error function

$$\mathrm{erfc}(\zeta) = 1 - \frac{2}{\sqrt{\pi}} \int_{0}^{\zeta} e^{-\sigma^2} d\sigma.$$

Consider those zeros $\zeta_k$ of erfc$(\zeta)$ which lie in the halfplane Im $\zeta > 0$ and arrange them as a sequence

$$\zeta_1, \zeta_2, \zeta_3, \cdots \qquad (|\zeta_1| \leq |\zeta_2| \leq |\zeta_3| \leq \cdots). \tag{24}$$

It is clearly possible to deduce from Theorem 1 the existence of an infinite sequence M of positive increasing integers

$$M: \quad m_1, m_2, \ldots, m_k, \cdots \tag{25}$$

and an associated positive, strictly increasing sequence $(R_m)$ $(m \in M)$ such that

$$m = [a(R_m)], \quad R_m \notin E, \quad R_m \to +\infty \qquad (m \to +\infty, \ m \in M). \tag{26}$$

Let $\phi_m(\zeta)$ be the function obtained by replacing, in the lefthand side of (19), R by $R_m$.

Then, the uniform convergence in (19), and Hurwitz's theorem, show that $\phi_m(\zeta)$ has zeros

$$\zeta_{m1}, \zeta_{m2}, \zeta_{m3}, \cdots$$

A. EDREI

such that

$$\zeta_{mk} \to \zeta_k \qquad (m \to +\infty,\ m \varepsilon M,\ k = 1,2,3,\ldots).$$

Hence, for  m large and $m \varepsilon M$, the partial sum $s_m(z)$ has zeros at the points

$$R_m \{1+(\zeta_k+\eta_{mk})\,(\frac{2}{b(R_m)})^{1/2}\} \qquad (k=1,2,3,\ldots), \tag{27}$$

with

$$\eta_{mk} \to 0 \qquad (m \to +\infty,\ m \varepsilon M;\ k=1,2,3,\ldots). \tag{28}$$

The presence and behavior of the conjugate zeros $\overline{\zeta}_{mk}$ is obtained by symmetry and requires no further elaboration.

In order to illustrate the meaning of our approximation (27) we evaluate asymptotically the quantities

$$R_m,\ b(R_m),\ a(R_m) \qquad (m \to +\infty)$$

associated with

$$f(z) = e_k(z) \qquad (k \geq 1).$$

The definitions yield, at once

$$e_k'(z) = \prod_{\ell=1}^{k} e_\ell(z),$$

$$a(r) = \prod_{\ell=1}^{k-1} e_\ell(r), \qquad \frac{a'(r)}{a(r)} = \sum_{\ell=0}^{k-1} \frac{e_\ell'(r)}{e_\ell(r)}, \tag{29}$$

$$\frac{b(r)}{a(r)} = 1\ (k=1), \qquad \frac{b(r)}{a(r)} = \frac{ra'(r)}{a(r)} \sim \prod_{\ell=0}^{k-2} e_\ell(r) \qquad (k \geq 2,\ r \to +\infty). \tag{30}$$

From (26) and (29) we deduce

$$a(R_m) \sim m, \tag{31}$$

and more precisely

$$\log m + 0(\frac{1}{m}) = \log R_m + R_m + \ldots + e_{k-2}(R_m) \qquad (m \to +\infty,\ m \varepsilon M). \tag{32}$$

Introducing the iterated logarithms

$$\log_1 r = \log r, \qquad \log_k r = \log(\log_{k-1} r) \qquad (k \geq 2),$$

we find

$$e_{k-2}(R_m) \sim \log m, \qquad e_{k-j}(R_m) \sim \log_{j-1} m \qquad (j=2,3,\ldots,k), \tag{33}$$

$$R_m \sim \log_{k-1} m \qquad (k \geq 2,\ m \to +\infty,\ m \varepsilon M). \tag{34}$$

Finally, combining (30), (31) and (33) we obtain

$$b(R_m) \sim m \log m (\log_2 m) \ldots (\log_{k-1} m) \qquad (k \geq 2). \tag{35}$$

The relations (34) and (35), used in (27), lead to approximations of those zeros of $s_m(z)$ which lie near the positive axis. It is not difficult to improve the crude approximation derived from (34). For instance, for $e_2(z)$ we find

$$R_m = \log m - \log \log m + 0\left(\frac{\log_2 m}{\log m}\right) \ (m \to +\infty), \tag{36}$$

and similar improvements may be used in the study of $e_k(z)$ $(k \geq 3)$. By analyzing the definition of $e_k(z)$ it is easy to prove that, if $f(z)$ is any-one of the functions $e_k(z)$ $(k \geq 1)$, the growth of the associated function $a(r)$ is so regular that the introduction, in (19), of an exception set of E is no longer needed. This fact has an interesting consequence:  the approxima-tion described in (27) holds for all large values of m and not merely for $m \varepsilon M$.

To complete the study of the partial sum $s_m(z)$ we need an extension of Theorem 1 valid along rays:

$$re^{i\varphi} \qquad (r>0,\ 0< \varphi \leq \pi,\ \varphi \text{ constant}). \tag{37}$$

Such an extension requires some additional information regarding the behavior of $f(z)$ along rays other than the positive axis; it will permit us to extend, to all the iterated exponentials $e_k(z)$ $(k \geq 2)$, the precise results obtained by Szegö for $\exp(z)$. The proofs are too long to be presented here; they will appear in a forthcoming paper [5].

1.  <u>Notational conventions.</u> The symbols F and E denote measurable subsets of the intervals $(1, +\infty)$. We always assume that F is of finite Lebesgue measure and E of finite logarithmic measure, that is

$$\int_E \frac{dx}{x} < + \infty.$$

We make little use of the symbols $o$ and $O$ which often mask some uniformities essential in our proofs. We use instead the symbols $\eta(r)$ to denote a positive function of the positive variable $r$ such that

$$\eta(r) \to 0 \qquad (r \to +\infty).$$

It is understood that $\eta(r)$ depends neither on $\Theta = \arg z$ nor on the auxiliary variables $w$ and $\zeta$ which appear in this paper.

By $\omega$ we mean a complex quantity such that

$$|\omega| \leq 1;$$

$\omega$ may be a function of all the variables and parameters of the problem.

We denote by $K$ a positive constant which may depend on all our parameters.

Inequalities such as

$$r > r_0, \quad x > x_0, \quad m > m_0, \quad \cdots$$

immediately following some relation mean that it is possible that the relation in question only holds for sufficiently large values of $r$, $x$, $m$, $\ldots$.

The symbols $\eta$, $K$, $r_0$, $x_0$, $m_0$, $\ldots$ may have different values in different places. Similarly $E$ is not necessarily the same set at each occurrence.

By $[X]$ we denote the greatest integer $\leq X$.

2. <u>Fundamental decomposition and estimates.</u> Let $f(z)$ given by its expansion (1), satisfy all the conditions of Theorem 1.

Let $R > R_0 > 0$ and define the positive function $m(R)$ by (18) and (5).

It is clear that $m(R)$ is a positive, nondecreasing, unbounded, integer-valued step-function.

Introduce an auxiliary complex variable $w$ and consider the expressions

$$U_m(w) = \frac{f(Rw)}{a_m R^m w^m} \qquad (w \neq 0), \tag{2.1}$$

$$Q_m(w) \frac{s_m(Rw)}{a_m R^m w^m} = \sum_{j=0}^{m} c_{-j}(m) w^{-j}, \tag{2.2}$$

$$G_m(w) = \sum_{j=1}^{\infty} \frac{a_{m+j}}{a_m} R^j w^j = \sum_{j=1}^{\infty} c_j(m) w^j, \tag{2.3}$$

where

$$c_j(m) = \frac{a_{m+j}}{a_m} R^j \qquad (j \geq -m). \tag{2.4}$$

From (2.1), (2.2) and (2.3) it follows that

$$U_m(w) = Q_m(w) + G_m(w) \qquad (w \neq 0).$$ 
(2.5)

The property III of admissible functions asserts that there are at most finitely many negative coefficients $a_j$. Hence

$$\sum_{j=0}^{\infty} |a_j| R^j = f(R) + 2 \sum_{a_j < 0} |a_j| R^j,$$
(2.6)

which, in view of (6), (13) and (18) implies

$$\sum_{j=0}^{\infty} |a_j| R^j \sim a_m R^m \{2\pi b(R)\}^{1/2} \qquad (R \to +\infty, \ m = m(R)).$$
(2.7)

We note that, for j fixed, positive, negative or zero, (2.4) and (13) yield

$$c_j(m) \to 1 \qquad (R \to +\infty, \ m = m(R) \to +\infty).$$
(2.8)

Since R and m(R) tend simultaneously to $+\infty$, explicit reference to R is unnecessary and we may state the validity of (2.8) under the assumptions: $m \to +\infty$ (m$\in$M), where M is a suitable infinite sequence of positive, increasing integers. The introduction of M may be necessary because certain inequalities, which appear later in our proof, are only valid if R$\notin$E (E of finite logarithmic measure). The proof of (2.8) also shows that, for all j,

$$|c_j(m)| < 2 \qquad (m > m_0, m_0 \text{ independent of j}).$$
(2.9)

From (2.2), (2.3), and (2.7) we deduce

$$|Q_m(w)| \leq \sum_{j=0}^{m} |c_{-j}(m)| |w|^{-j} \leq 2\{2\pi b(R)\}^{1/2} \qquad (|w| \geq 1, \ m > m_0, \ m \in M),$$
(2.10)

and

$$|G_m(w)| \leq \sum_{j=1}^{\infty} c_j(m) |w|^j \leq 2\{2\pi b(R)\}^{1/2} \qquad (|w| \leq 1, \ m > m_0, \ m \in M).$$
(2.11)

3. <u>Taylor's formula for</u> log f(z) <u>and estimate of</u> $U_m(w)$. The following elementary consequence of Cauchy's formula is almost immediate.

<u>Lemma 3.1.</u> <u>Suppose that</u> f(z) <u>is regular, that it has no zeros in this disk</u>

$$|z - z_0| \leq 2H|z_0| \qquad (z_0 \neq 0, \ 0 < H),$$
(3.1)

<u>and let</u> log f(z) <u>be the branch of the logarithm, regular in this disk,</u> <u>reducing to some determination</u> log f($z_0$) <u>at</u> z=$z_0$.

Denote by s an auxiliary complex variable, confined to the disk

28                                    A. EDREI

$$|s| \leq \frac{H}{2} |z_0|. \tag{3.2}$$

<u>Then</u>

$$\log f(z_0+s) - \log f(z_0) = s\frac{f'(z_0)}{f(z_0)} + \frac{s^2}{2}\left(\frac{f''(z_0)}{f(z_0)} - \left\{\frac{f'(z_0)}{f(z_0)}\right\}^2\right) + \epsilon(z_0,s), \tag{3.3}$$

<u>with</u>

$$|\epsilon(z_0,s)| \leq 4 \left(\frac{|s|}{H|z_0|}\right)^3 \max_{0 \leq \theta \leq \pi} \overset{+}{\log}\left|\frac{f(z_0(1+2He^{i\theta}))}{f(z_0)}\right|. \tag{3.4}$$

<u>Proof.</u> Put

$$g(s) = \log\left(\frac{f(z_0+s)}{f(z_0)}\right) \quad (g(0) = 0).$$

By Cauchy's formula

$$g(s) = \frac{1}{2\pi i}\int_{C_1} \frac{g(\zeta)}{\zeta-s} d\zeta \quad (|s| \leq \frac{H}{2} |z_0|), \tag{3.5}$$

where

$$C_1: \zeta = Hz_0 e^{i\theta} \quad (0 \leq \theta \leq 2\pi).$$

By the elements of function theory we conclude that (3.3) holds with

$$\epsilon(z_0,s) = \frac{s^3}{2\pi i}\int_{C_1} \frac{g(\zeta)}{\zeta^3(\zeta-s)} d\zeta. \tag{3.6}$$

By the Borel-Carathéodory inequality [12; p. 175] we see that

$$\max_{|\zeta|=H|z_0|} |g(\zeta)| \leq 2 \max_{|\zeta|=H|z_0|} \text{Re}\{g(2\zeta)\} = 2 \max_{0 \leq \theta \leq 2\pi} \overset{+}{\log}\left|\frac{f(z_0(1+2He^{i\theta}))}{f(z_0)}\right|.$$

Hence the estimate (3.4) follows at once from (3.6).

We now apply the above lemma to the function f(z) with

$$z_0=R, \quad 2H= \{a(R)\}^{-\alpha}, \quad s= \omega BR\{b(R)\}^{\frac{1}{2}}, \tag{3.7}$$

where B>0 is a given constant (as large as we please).

Our choice of R will be restricted by the condition (3.12) stated below.  Hence with H and $z_0$ given by (3.7), the disk in (3.1) will, by assumption, contain no zeros of f(z).

    From (16) and (22) we deduce

$$2b(r) > \{a(r)\}^{1-(1/\beta)} \qquad (r>r_0),$$  (3.8)

**and therefore**

$$\frac{|s|}{H|z_0|} < 2B\{a(R)\}^{\alpha}\{b(R)\}^{-\frac{1}{2}} < 2^{\frac{3}{2}}Ba^{\alpha-\frac{1}{2}+\frac{1}{2\beta}} \qquad (R>R_0),$$

with

$$\alpha-\frac{1}{2}+\frac{1}{2\beta} < \frac{\alpha-1}{3} < 0.$$

This shows that the condition (3.2) is satisfied if R is large enough and
hence Lemma 3.1 yields

$$\left|\varepsilon(z_0,s)\right| \leq 0(\{a(R)\}^{\alpha-1})\{\log M(R+Ra(R)^{-\alpha})-\log f(R)\} \qquad (R \to+\infty).$$  (3.9)

In view of (5) and (12)

$$\log M(R + R\{a(R)\}^{-\alpha}- \log f(R)=o(1) + \int_{R}^{R+R\{a(R)\}^{-\alpha}} \frac{a(t)}{t}dt \quad (R\to+\infty).$$  (3.10)

Combining (3.9) and (3.10) we find

$$\left|\varepsilon(z_0,s)\right| \leq o(\{a(R)\}^{-1})a(R+R\{a(R)\}^{-\alpha}) \qquad (R \to+\infty),$$  (3.11)

where, as a simple consequence of Borel's growth lemma,

$$a(R+R\{a(R)\}^{-\alpha}) <2a(R) \qquad (R\notin E).$$  (3.12)

Taking (3.12) for granted we see that

$$\varepsilon(R,s) \to 0, \qquad (R\to+\infty,\ R\notin E)$$  (3.13)

uniformly for all

$$|s| \leq BR\{b(R)\}^{-\frac{1}{2}}.$$  (3.14)

For the sake of completeness we indicate how (3.12) follows from one
of the standard forms of Borel's lemma [1; pp. 15-16]:

If T(x)>1 is non-decreasing and unbounded then

$$\log T(x+ \frac{1}{\log T(x)}) < 2^{\alpha}\log T(x) \qquad (x>0,\ x\notin F,\ \alpha >0),$$  (3.15)

where F is a measurable subset of $(1,+\infty)$, of finite measure. [The positive
constant $\alpha$ is arbitrary.]

If V(x) is a given, positive, increasing, unbounded function of x>0,
we apply (3.15) to the function T(x) defined by

$$V(x) = \{\log T(x)\}^{1/\alpha}$$

and find

$$V(x + \frac{1}{\{V(x)\}^{\alpha}}) < 2V(x) \qquad (x>0, \ x\not\in F, \ \alpha>0). \tag{3.16}$$

The latter inequality, applied to

$$V(x) = a(\exp(x)) \qquad (\exp(x)=R),$$

yields

$$a(\exp(x+ \frac{1}{\{a(R)\}^{\alpha}})) < 2a(R) \qquad (R\not\in E), \tag{3.17}$$

and the proof of (3.12) is completed by noticing that

$$R(1+ \frac{1}{\{a(R)\}^{\alpha}}) < \exp(x)\exp( \frac{1}{(a(R))^{\alpha}}).$$

The condition $R\not\in E$ expresses in terms of the variable R, the condition $x\not\in F$.

We state, in the following Lemma 3.2, a summary of the results of this section.

Lemma 3.2.   <u>Let the assumptions and notations of Theorem 1 be unchanged.
Let w be an auxiliary complex variable restricted by the condition</u>

$$|w-1| \leq B\{b(R)\}^{-\frac{1}{2}}. \tag{3.18}$$

<u>Then Lemma 3.1 and the notational conventions yield, uniformly in w,</u>

$$\log f(Rw) = \log f(R) + (w-1)a(R) + \frac{(w-1)^2}{2}(b(R)-a(R)) + \omega\eta(R) \qquad (R\to+\infty, \ R\not\in E),$$

$$U_m(w) = (2\pi b(R))^{1/2}\exp\{(w-1)m + \frac{(w-1)^2}{2}(b(R)-m)) - m\log w + \omega\eta(R)\} \tag{3.19}$$

$$(m=m(R) = [a(R)], \ R\to+\infty, \ R\not\in E).$$

Using for $\log w$ the approximation

$$\log w = (w-1) - \frac{1}{2}(w-1)^2 + \omega|w-1|^3 \qquad (|w-1| \leq \frac{1}{2}),$$

and taking (3.8) and (3.18) into account, we find

$$m|w-1|^3 = O(\{a(R)\}^{-\frac{1}{2} + \frac{3}{2\beta}}) \qquad (R\to+\infty).$$

The inequality for $\beta$ in (16) now shows that

$$-\frac{1}{2} + \frac{3}{2\beta} < -2\alpha$$

and consequently

$$m|w-1|^3 \leq \eta(R) \qquad (m=m(R))$$

under the restriction (3.18).

Returning to (3.19) we obtain

Lemma 3.3. Let $\zeta$ be an auxiliary complex variable and let $f(z)$ satisfy the condition of Theorem 1. Then, as

$$R \to +\infty, \quad R \notin E, \quad m=m(R), \tag{3.20}$$

we have

$$\{2\pi b(R)\}^{-\frac{1}{2}} U_m(1+\zeta\{\tfrac{2}{b(R)}\}^{1/2}) \to \exp(\zeta^2), \tag{3.21}$$

uniformly on every compact subset of the $\zeta$-plane.

4. The left-hand side of (19) is uniformly bounded. An inspection of (2.2) and (2.10) reveals that, under the assumptions (3.20), the functions

$$\phi_m(\zeta) = \frac{s_m(R(1+\zeta\{\tfrac{2}{b(R)}\}^{1/2}))}{f(R)(1+\zeta\{\tfrac{2}{b(R)}\}^{1/2})^m} = Q_m(1+\zeta\{\tfrac{2}{b(R)}\}^{1/2})^m (2\pi b(R))^{-\frac{1}{2}}$$

are uniformly bounded on the set

$$\{\zeta : |\zeta| \leq B, \ |1+\zeta\{\tfrac{2}{b(R)}\}^{1/2}| \geq 1\}.$$

On the set

$$\{\zeta : |\zeta| \leq B, \ |1+\zeta\{\tfrac{2}{b(R)}\}^{1/2}| \leq 1\},$$

the boundedness of $\{\phi_m(\zeta)\}_m$ follows from (2.5), (2.11) and (3.21).

Hence the family $\{\phi_m(\zeta)\}_m$ $(m \in M)$ is bounded on any compact subset of the $\zeta$-plane.

In view of Vitali's theorem, it suffices, in order to complete the proof of Theorem 1, to evaluate $\lim \phi_m(t)$ for $t$ real, $1<t<2$.

5. Determination of $\lim \phi_m(t)$. We now consider the coefficients $c_j(m)$ introduced in (2.4) and make full use of Hayman's relation (13) and of our notational convention regarding $\omega$ and $\eta(r)$.

With

$$m=m(R) = [a(R)] = a(R) + \omega \tag{5.1}$$

32           A. EDREI

we find, uniformly for all integers $j$,

$$c_j(m) = \exp\left(-\frac{(j+\omega)^2}{2b(R)}\right) + \omega\eta(R) \qquad (R\to+\infty,\ R\notin E).\tag{5.2}$$

To simplify (5.2) need upper and lower bounds for $b(R)$ in terms of $m$. Starting from (3.8) (and in view of the inequality for $\beta$ in (16)), we find

$$b(R) > \frac{1}{2}\{a(R)\}^{\frac{2}{3}+\mu} \qquad (\mu=\frac{4\alpha}{3} > 0),$$

which implies

$$b(R)m^{-\frac{2}{3}} > \frac{1}{2}m^{\mu} \qquad (m=m(R),\ R\to+\infty,\ R\notin E).\tag{5.3}$$

In the opposite direction (21) implies

$$b(R)m^{-\frac{5}{4}} \to 0 \qquad (R\to+\infty,\ R\notin E).\tag{5.4}$$

Define

$$L = L(m) = [m^{2/3}].$$

Then, for $|j|\leq L(m)$, (5.2) and (5.3) imply

$$c_j(m) = \exp\left(-\frac{j^2}{2b(R)}\right) + \omega\eta(R) \qquad (m=m(R),\ R\to+\infty,\ R\notin E).\tag{5.5}$$

For $|j|>L(m)$, (5.4) yields

$$\frac{(j+\omega)^2}{2b(R)} > m^{1/2} \qquad (m=m(R),\ R\to+\infty,\ R\notin E)$$

and hence (5.2) takes the form

$$c_j(m) = \omega\eta(R) \qquad (m=m(R),\ R\to+\infty,\ R\notin E).\tag{5.6}$$

Using (5.5) and (5.6) in (2.2) we find

$$Q_m(w) = \sum_{j=0}^{L(m)} \exp\left(-\frac{j^2}{2b(R)}\right)w^{-j} + \omega\eta(R)\sum_{j=0}^{m}\frac{1}{|w|^j} \qquad (w\neq 0).\tag{5.7}$$

We now set in (5.7)

$$w = 1+t\left\{\frac{2}{b(R)}\right\}^{1/2} \qquad (1\leq t\leq 2);$$

this substitution leads to

$$Q_m\left(1+t\left\{\frac{2}{b(R)}\right\}^{1/2}\right) = \sum_{j=0}^{L(m)}\exp\left(-\frac{j^2}{wb(R)}\right)\left(1+t\left\{\frac{2}{b(R)}\right\}^{1/2}\right)^{-j} + \omega\eta(R)\{b(R)\}^{1/2}.\tag{5.8}$$

Comparing (5.8) and (4.1) we see that Theorem 1 will be proved as soon as we establish the existence and find the value of

$$\lim_{\substack{R \to +\infty \\ R \notin E}} (2\pi b(R))^{-\frac{1}{2}} \sum_{j=0}^{L(M)} \exp\left(-\frac{j^2}{2b(R)}\right)(1+t\,\{\tfrac{2}{b(R)}\}^{1/2})^{-j} = \tag{5.9}$$

$$\lim_{\substack{m \to +\infty \\ m \in M}} \Phi_m(t) \qquad (1 \le t \le 2,\ m = [a(R)]).$$

6. <u>Comparison with an integral</u>. To simplify the writing we set

$$\gamma = \{2b(R)\}^{-(1/2)}, \tag{6.1}$$

so that the sum in (5.9) may be rewritten as

$$Y_m(t) = \sum_{j=0}^{L(m)} \exp(-j^2\gamma^2)(1+2t\gamma)^{-j}. \tag{6.2}$$

Using the approximation

$$\log(1+u) = u+\omega u^2 \qquad (|u| \le \tfrac{1}{2})$$

and the fact that, by (5.3)

$$\frac{j}{b(R)} = 2j\gamma^2 \le 2m^{-\mu} \qquad (0 \le j \le m^{2/3},\ m > m_0),$$

we see that

$$Y_m(t) = X_m(t) + \omega K m^{-\mu} X_m(t) \qquad (1 \le t \le 2,\ m > m_0) \tag{6.3}$$

where

$$X_m(t) = \sum_{j=0}^{L(m)} \exp(-j^2\gamma^2 - 2jt\gamma) = \exp(t^2) \sum_{j=0}^{L(m)} \exp(-(j\gamma+t)^2). \tag{6.4}$$

From the elementary inequalities

$$\exp(-\{(j+1)\gamma+t\}^2) < \int_j^{j+1} \exp(-(x\gamma+t)^2)\,dx < \exp(-\{j\gamma+t\}^2)$$

we conclude that

$$0 < \exp(-t^2) X_m(t) - \int_0^{L(m)+1} \exp(-(x\gamma+t)^2)\,dx < \exp(-t^2),$$

$$0 < X_m(t) - \exp(t^2) \int_0^{\infty} \exp(-(x\gamma+t)^2)\,dx + J_m(t) < 1, \tag{6.5}$$

with

$$J_m(t) = \exp(t^2) \int_{L(m)+1}^{+\infty} \exp(-(x\gamma+t)^2)\,dx \qquad (1 \le t \le 2).$$

Obviously

$$0 < J_m(t) < \exp(-m^{4/3}\gamma^2) \int_{m^{2/3}}^{+\infty} \exp(-2x\gamma)\,dx < (2\gamma)^{-1} \exp(-m^{4/3}\gamma^2). \tag{6.6}$$

34                                    A. EDREI

By (6.1) and (5.4)

$$\gamma^2 = \frac{1}{2b(R)} > \frac{1}{2} m^{-\frac{5}{4}}, \quad \frac{1}{\gamma} < 2^{1/2} m^{5/8}.$$

Hence (6.6) yields

$$0 < J_m(t) < m^{5/8} \exp(-m^{(1/12)}) \to 0 \quad (m \to +\infty). \tag{6.7}$$

Combining (6.1), (6.3), (6.5), (6.7), and using an obvious change of
variable in the integral, we obtain

$$0 < Y_m(t)(1+\omega Km^{-\mu})^{-1} - \{2b(R)\}^{1/2} \exp(t^2) \int_t^{+\infty} e^{-\sigma^2} d\sigma < 2. \tag{6.8}$$

Since

$$\int_t^{+\infty} \exp(-\sigma^2) d\sigma = \frac{1}{2}\sqrt{\pi} - \int_0^t \exp(-\sigma^2) d\sigma,$$

(6.8) implies

$$\lim_{\substack{R \to +\infty \\ R \notin E}} (2\pi b(R))^{-\frac{1}{2}} Y_m(t) = \frac{1}{2} e^{t^2} (1 - \frac{2}{\sqrt{\pi}} \int_0^t \exp(-\sigma^2) d\sigma). \tag{6.9}$$

We have already observed that, in view of Vitali's convergence theorem,
(6.9) establishes Theorem 1.

7. __Proof of Remark 4.__ Since f(z), defined by (23), has no zeros, the disks
S in (15) play no role. All but a finite number of coefficients of the
Taylor expansion of g(z) are positive, because g(z) is admissible by
assumption. Consequently

$$g(x) > 0, \quad g^{(k)}(x) > 0 \quad (k=1,2,3; \ x > x_0). \tag{7.1}$$

Expressed in terms of g'(x) and g"(x), the functions a(x) and b(x)
associated with f(z) are

$$a(x) = xg'(x) > x \quad (x > x_0), \tag{7.2}$$

$$b(x) = xg'(x) + x^2 g''(x). \tag{7.3}$$

Now (16) follows from (7.2) and (17) is a consequence of (7.3) and (7.1).
We have thus proved Remark 4.

Acknowledgement

The research of the author was supported by a grant from the National
Science Foundation.

## References


1.  O. Blumenthal, Principes de la théorie des fonctions entieres d'ordre infini, Gauthier-Villars, Paris, 1910.

2.  F. Carlson, Sur les fonctions entières, C.R. Acad. Sci. Paris 179 (1924), 1583-1585.

3.  F. Carlson, Sur les fonctions entières, Ark. Mat. Aston. Fys. 35A (1948), 1-18.

4.  A. Edrei, Sections of the Taylor expansions of Lindelöf functions, to be published.

5.  A. Edrei, The conjecture of Saff-Varga for entire functions admissible in the sense of Hayman, to be published.

6.  A. Edrei, E.B. Saff and R.S. Varga, Zeros of the partial sums of Mittag-Leffler's expansion and of certain functions of genus zero, Lecture Notes in Mathematics, Springer-Verlag, New York, Heidelberg, Berlin, (1983).

7.  W.K. Hayman, A generalization of Stirling's formula, Journal f.d. reine und angew. Mathematik, 196 (1956), 67-95.

8.  G. Pólya and G. Szegö, "Aufgaben und Lehrsätze aus der Analysis," Vol. 2, Springer-Verlag, Berlin, 1925.

9.  P.C. Rosenbloom, "Sequences of polynomials, especially sections of power series," Ph.D. dissertation, Stanford, 1943.

10. P.C. Rosenbloom, Distribution of zeros of polynomials, in "Lectures on Functions of a Comples Variable" (W. Kaplan, Ed.), Univ. of Michigan Press, Ann Arbor, 1955.

11. G. Szegö, Über eine Eigenschaft der Exponentialreihe, Sitzungber. der Berl. Math. Ges. 23 (1924), 50-64.

12. E.C. Titchmarsh, "The Theory of Functions," 2nd ed., Oxford Univ. Press, London, New York, 1952.



Department of Mathematics
Syracuse University
Syracuse, New York

Contemporary Mathematics
Volume **25**, 1983

## FIXED POINTS OF COMPOSITE ENTIRE FUNCTIONS

### Fred Gross

In 1926 Fatou [1] stated that if $f(z)$ is any nonlinear entire function, then its iterate

$$f_2(z) = fof(z)$$

has a fix-point. This was subsequently proved by P. Rosenbloom [10] in 1952.

Rosenbloom proved this result using Picard's Theorem. Since it is rather simple and elegant it is worth reproducing here.

Suppose that $fof(z)$ has no fix-point. Then $f(z)$ has no fix-point either and thus,

$$F(z) = \frac{f(z)-z}{fof(z)-z}$$

has no zeros or poles. Hence, by Picard's Theorem it must be a constant C, say. Thus the above equation yields

$$C[fof(z)-z] = f(z)-z \;\&\; C \neq 1 \text{ otherwise } f(f(z))=f(z), \text{ a contradiction.}$$
Thus,

$$C[f'(z)f'(f(z))-1] = f'(z)-1$$

or

$$f'(z)[Cf'(f(z))-1] = C-1 \neq 0.$$

Thus, $f'(z)$ is never zero and hence must attain all non-zero values. But the last equation also implies that $f'(f(z)) \neq 1/C$, a contradiction. Thus $fof(z)$ must have a fix-point.

Rosenbloom [1] generalized this result using Nevanlinna Theory and proved Theorem A. If f and g are transcendental and entire, then f or f(g) must have infinitely many fix-points. Rosenbloom [10] also proved:

Theorem B. If $P(z)$ is a nonlinear polynomial and $f(z)$ is entire and transcendental, then $P(f(z))$ has infinitely many fix-points.

An independent more elementary proof of Theorem B was also given by the author and Yang [4]. Rosenbloom's result was generalized by the author [3] as follows:

Theorem C. Let $F(F*)$ denote the family of entire (meromorphic) functions which have at most a finite number of fix-points. Then:

    (i)    every entire function has at most one factorization $f(g(z))$, f

           transcendental, $f\varepsilon F$, and g entire, and

    (ii)   every meromorphic function has at most two distinct factorizations

           $f_i(g_i(z))$, $f_i$ meromorphic and not rational, $f_i\varepsilon F*$ and $g_i$ entire,

           i=1,2.

Theorem A is an immediate corollary of Theorem C.  We also have:

Corollary:  If f is transcendental and meromorphic and g and h are transcendental and entire, then one of f(z), f(g(z)) and f(g(h(z))) has infinitely many fix-points.

These results have lead the author to conjecture [2]:

Conjecture 1:  If f is any nonlinear meromorphic function and g is transcendental entire, then gof has infinitely many fix-points.

More specifically for entire functions the above conjecture becomes

Conjecture 2:  If f is any nonlinear entire function and g(z) is transcendental entire, then fog has infinitely many fix-points.

By virtue of Theorem B. conjecture 2 is true for, f a nonlinear polynomial.

Theorem C also shows that conjecture 2 is valid when f is periodic.

The author and Yang [4] proved that Conjecture 2 is valid when fog is of finite order.  Yang [12] later extended this result for f and g subject to a variety of growth conditions.

Conjecture 2 can be formulated equivalently as

Conjecture 2':  Let $Q(z)$ be any polynomial $(\not\equiv 0)$ and let $\alpha(z)$ be any non-constant entire function then $Q(z)e^{\alpha(z)} + z$ is prime.

In this formulation numerous partial solutions exist.  See for example [5], [6].

The most interesting result concerning Conjecture 2 that has been obtained thus far is a proof when f and g satisfy certain growth conditions. This result can be stated as follows:

Theorem D:  [7]  Let f and g be two entire functions both of which are non-linear.  Suppose that one of f and g is transcendental.  Suppose also that one of f and g is of finite order, while the other h say, satisfies $\lim_{r \to \infty} T(r,h)r^{-N}=0$ for some positive interger N.  (Here $T(r,h)$ is the Nevanlinna Characteristic function of h.).  Then $fog(z) = z$ has infinitely many solutions z.

We sketch the proof:  (For complete proof see [7])

Lemma 1: [4]  If f and g are nonlinear and one of the two functions f and g is transcendental while the other is a polynomial, then $fog(z) = z$ has infinitely many solutions z.

This follows from Theorem B.

Lemma 2.  Steinmetz [11].  Let $F_0, F_1, \ldots, F_m$ be not identically vanishing entire functions and let $h_0, h_1, \ldots, h_m$ be arbitrary meromorphic functions $(m \geq 1)$.  Let g be a nonconstant entire function, let K be a positive real number, and let $(r_j)_{j=1}^{\infty}$ be an unbounded monotone increasing sequence of positive real numbers such that, for each j,

$$T(r_j,h_n) \leq KT(r_j,g) \ (n=0,1,\ldots m),$$

while, also on $(r_j)_{j=1}^{\infty}$,

$$T(r,g') \leq (1+o(1))T(r,g).$$

If $F_i$ and $h_i$ $(i=0,1,\ldots,m)$ satisfy

$$F_0(g)h_0+F_1(g)h_1+\ldots+F_m(g)h_m=0$$

then there exist polynomials in z, $P_0$, $P_1$, ..., $P_m$ not all zero such that

$$P_0F_0+P_1F_1+\ldots+P_mF_m\equiv0.$$

The original result of Steinmetz is proved under the condition that

$$T(r,h_n) \leq KT(r,g)$$

holds for sufficiently large n.

A modification of Steinmetz's proof yields Lemma 2 in the stated form.

Lemma 3. [8]  If f(z) is entire and $a_i(z)$, i=1 or 2, are distinct entire functions satisfying $T(r, a_i(z))=o(T(r,f))$, then

$$(1+o(1))T(r,F)\leq \sum_{i=1}^{2} N(r,(F-a_i(z))^{-1})+o(\log r + \log T(r,F))$$

outside a set of finite measure.

Lemma 4. [7]  Suppose that f and g are entire functions at least one of which is transcendental.  Suppose that f is of finite order of growth while g satisfies $\lim_{r\to\infty} T(r,g)r^{-N}=0$ for some integer N.  Suppose that $f(g)=Qe^{\alpha}+z$ where Q is a polynomial and $\alpha$ is entire.  Then there exists an unbounded monotone increasing sequence of positive numbers $(r_j)_{j=1}^{\infty}$ such that

(i)  for some C > 0, $T(r_j,\alpha') < CT(r_j,g)$, and

(ii)  on the sequence $(r_j)_{j=1}^{\infty}$, $T(r,g') \leq (1+o(1))T(r,g)$.

The proof of this lemma is proved by mathematical induction and involves some simple growth arguments.

Lemma 5. [3]  If f(g) has only a finite number of fixed points then g(f) also has only a finite number of fixed points.

Lemma 6. [9]  If g is transcendental and $\psi_1$ and $\psi_2$ are relatively prime polynomials in g and z with $\psi_2\not\equiv0$, then $T(r,\psi_1/\psi_2) = (\max\{\deg_g\psi_1,\deg_g\psi_2\})T(r,g) +$

40                             F. GROSS

$o(R(r,g))$.

We are now ready to complete the proof of the theorem.

Sketch of proof of Theorem:  Suppose f and g satisfy the hypothesis of Theorem D but that fog has at most finitely many fixed points.  Then we may write

$$f(g(z))-z=Qe^{\alpha} \tag{1}$$

where Q is a polynomial and $\alpha$ is an entire function.

By virtue of Lemma 5 we may assume that f is of finite order and by virtue of Lemma 1 f and g may both be assumed to be transcendental. Equation. (1) yields

$$g'f'(g)=(Q'+\alpha'Q)e^{\alpha}+1 \tag{2}$$

or

$$g'f'(g)=(\frac{Q'}{Q}+\alpha')(f(g)-z)+1. \tag{3}$$

By Lemma 4, on $\{r_j\}_{j=1}^{\infty}$

$$T(r,\alpha')=o(T(r,g)) \quad \text{and} \quad T(r,g')=(1+o(1)T(r,g)$$

while $\gamma = \frac{Q'}{Q} + \alpha'$ satisfies $T(r,\gamma)=o(T(r,g)$

on the same sequence $\{r_j\}_{j=1}^{\infty}$.

Now applying our modified version of Steinmetz's lemma, Lemma 2 to equation (3), where $F_0$, $F_1$ and $F_2$ are f, f' and 1, respectively, we obtain

$$\phi_1 f' - \phi_2 f - \phi_3=0, \tag{4}$$

where $\phi_1, \phi_2,$ and $\phi_3$ are relatively prime polynomials in z that are not all zero.

If $\phi_1 \equiv 0$, then f is a rational function; hence $\phi_1 \equiv 0$.  Substituting g for z in equation (4) yields

$$\phi_1(g)f'(g)=\phi_2(g)f(g)+\phi_3(g). \tag{5}$$

Multiply (5) by g' and (3) by $\phi_1(g)$ yields

$$\phi_1(g)(\gamma(f(g)-z)+1) =g'(\phi_2(g)(f)(g)+\phi_3(g)), \tag{6}$$

where $\gamma$ recall is $(Q'/Q + \alpha')$ but

$$H=\phi_1(g)\gamma - g'\phi_2(g) \equiv 0 \tag{7}$$

and

$$K=\phi_1(g)(1-z) - g'\phi_3(g) \equiv 0, \tag{8}$$

for otherwise solving equation (6) for f(g) yields f(g) = -H/K which implies that $T(r,f(g))=0(T(r,g))$ as $r\to\infty$ on the sequence $r_j$. This is impossible, since f and g are both transcendental. Equations (7) and (8) yields

$$\phi_1(g) = (z\phi_2(g) + \phi_3(g))g' = \phi_1(g). \tag{9}$$

Since $\phi_1(g) \neq 0$, we have $z\phi_2(g) + \phi_3(g) \neq 0$. Since $T(r,g') = T(r,g) + o(T(r,g))$, except for a set of finite measure we have from equation (9) via Lemma 6 that deg $\phi_i(g) \leq 1$ for $i=1,2,3$. Thus, equation (7) above i.e.,

$$\phi_1(g)\gamma = g'\phi_2(g) \tag{10}$$

implies that

$$(ag+b)\gamma = g'(cg+d) \tag{11}$$

for constants a, b, c and d with not both a and b being zero.

Now $\gamma(\frac{a'}{Q} + \alpha')$ has only finitely many poles, thus, for $\alpha\neq0$ equation (11) yields that ag+b can have only finitely many zeroes. Thus,

$$g=pe^{\eta}-ba^{-1}$$

where $\eta=\eta(z)$ is entire and p is a polynomial. When a=0 then equation (11) yields that $\gamma = \frac{Q'}{Q} + \alpha'$ must be entire and thus, Q must be a constant. We may assume that this constant in fact is unit and the equation (11) yields

$$b\alpha' = g'(g+d). \tag{13}$$

So that $\alpha=\Theta(g)$, where $\Theta$ is a polynomial. In this latter case, using the equation

$$f(g)=e^{\alpha} +z \tag{14}$$

We have $f(g) - e^{\Theta(g)}=z$, which implies that g is a linear function, contrary to the hypotheses.

In the former case, equation (12) yields

$$g(f)=p(f)e^{\eta(f)}-ba^{-1}. \tag{15}$$

Since f(g) has at most finitely many fix points, by hypothesis, by virtue of Lemma 5 the same is true of g(f). Thus, we have

$$g(f)=p(f)e^{\eta(f)}-ba^{-1}=Le^{\beta}+z, \tag{16}$$

where L is a polynomial and $\beta$ is entire: By virtue of Lemma 3 equation (16) is impossible. This concludes our proof.

Conjecture 2 for the general case with no growth assumptions remain open and in fact seems to be rather difficult to prove. Even the case where $g(z)=$

f(az) for some constant a different then 0 or 1 remains an open question.
This question can be stated as:  Does fof(z)-Cz have infinitely many zeroes
for all C$\neq$0.

Concluding Remarks:  As we stated earlier, Conjecture 2 can be stated in the
form of a factorization problem that is:  For any polynomial Q(z) and any
entire function $\alpha(z)$, $Q(z)e^{\alpha(z)} + z$ is prime.  Furthermore, it should be noted
that Theorem C which generalizes fix-point Theorems of Rosenbloom and Fatou
is stated as a Theorem on factorization.  Steinmetz's Theorem is also a
generalized factorization result.  Numerous other results can be formulated
in factorization terminology. Looking at problems such as the fix-point
problem for composite functions as factorization problems has certain
advantages, among them the fact that partial results otherwise not attainable
can be deduced and through these results in turn one gains deeper insights
into some of the difficulties involved.  Some critics of the subject have
remarked that results on factorizations are somewhat fragmented and that
the no complete theory seems to be forthcoming.  Well, indeed it is a bit
unrealistic to expect a "prime function theorem" or an analogue of number
theory.  The subject matter is much too difficult for that.  Nevertheless
some very interesting and difficult results have been obtained on factorization and it is hoped that this paper will inspire the reader to delve more
deeply into the subject.

## References


1.  P. Fatou, Sur literation des fonctions transcendantes entières, Acta
    Mathematica Vol. 47, 1926.

2.  F. Gross, Factorization of mermorphic functions, U.S. Government Printing
    Office, Washington, D.C., 1972.

3.  F. Gross, Factorization of mermorphic functions, Trans. Amer. Math. Soc.
    131 (1968) pp 215-272.

4.  F. Gross and C.C. Yang, The Fix-points and Factorization of Mermorphic
    Functions, Trans. Amer. Math. Soc. (1968) 1972.

5.  F. Gross, Prime Entire Functions, Trans. AMS 161 (1971, 219-233.

6.  F. Gross, C. Osgood and C.C. Yang, Prime Entire Functions with Prescribed
    Nevanlinna Deficiency, Nagoya Math. J. 47 (1972), 91-99.

7.  F. Gross and C. Osgood, On Fixed Points of Composite Entire Functions,
    Proc. London Math. Soc. to appear.

8.  W.K. Hayman, Meromorphic Functions, Oxford Monographs, Clarendon Press,
    Oxford, (1964).

9.  A.Z. Mohonko, The Nevanlinna Characteristics of Certain Meromorphic
    Functions, Teor Funkcii Funkcional, Anal. Prizen, 14 (1971), 83-87.

10. P. Rosenbloom, The Fix-points of Entire Functions, Medd. Lunds Univ.
    Mat. Sem. Suppl.-Bd. M. Riesz, 186 (1952).

11. M. Steinmetz, Über die Faktorisierbaren Lösungengewohnlicher
    Differetialgleichungen, Math. Z., 170 (1980), 169-180.

12. C.C. Yang, Further results on the Fix-points of Composite Transcendental
    Entire Functions, J. of Math. Anal. and Applic., Vol. 90, 1, Nov. 1982.



Department of Mathematics
University of Maryland at Baltimore County
Baltimore, Maryland  21228

Contemporary Mathematics
Volume 25, 1983

## THE GROWTH OF ALGEBROID FUNCTIONS
## WITH SEVERAL DEFICIENT VALUES

Yong-Xing Gu

1.  Edrei and Fuchs [1] proved that the lower order of meromorphic functions with two deficient values is positive.  Ozawa [2] proved that the lower order of a n-valued integral algebroidal function with n finite deficient values is also positive.  For a n-valued meromorphic algebroidal function with n + 1 deficient values, after adding a condition Ozawa [2] showed also that its lower order is positive.  At the time, he conjectured that this condition may be eliminated.  This paper justifies his conjecture.

2.  Let $w(z)$ be a n-valued meromorphic algebroidal function in $|z| < + \infty$, defined by an irreducible equation

$$A_n(z)w^n + A_{n-1}(z)w^{n-1} + \ldots + A_o(z) = 0, \tag{1}$$

where $A_n(z)$, $A_{n-1}(z)$, $\ldots$, $A_o(z)$ are entire functions without any common zero.

Let $T(r)$ be the characteristic function of $w(z)$, and set

$$A(z) = \max_{o \le i \le n} |A_i(z)|$$

$$\mu(r, A) = \frac{1}{2\pi n} \int_o^{2\pi} \log A(re^{i\theta}) d\theta.$$

By a theorem of Valiron [3] we have

$$|T(r) - \mu(r, A)| < K \tag{2}$$

From now on, K denotes a positive constant.

3.  We come to prove the following

Theorem 1.  Let $w(z)$ be a n-valued transcendental algebroidal function in $|z| < + \infty$ with n + 1 deficient values $a_i$ (i = 1, 2, $\ldots$, n + 1).  Then the lower order of $w(z)$ is positive.

Proof.  Without loss of generality we may assume that $a_1$, $a_2$, $\ldots$, $a_n$ are finite and $a_{n+1}$ is finite or infinite.

Put

$$g_i(z) = A_n(z)a_i^n + A_{n-1}(z)a_i^{n-1} + \ldots + A_1(z)a_i + A_o(z) \quad (i = 1, 2, \ldots, n)$$

$$g_{n+1}(z) = \begin{cases} A_n(z)a_{n+1}^n + \ldots + A_1(z)a_{n+1} + A_o(z), & \text{if } a_{n+1} \ne \infty, \\ A_n(z), & \text{if } a_{n+1} = \infty. \end{cases}$$

46                              Y. GU

Let $\alpha_\lambda$ ($\lambda = 1, 2, \ldots$) be zeros of $g_{n+1}(z)$ and $\beta_\mu^{(i)}$ ($\mu = 1, 2, \ldots$) be poles

of $\dfrac{g_i(z)}{g_{n+1}(z)}$ ($i = 1, 2, \ldots, n$).

By (7.4) in [1], for $|z| = r$, $R = \sigma r$, $\sigma > 1$ we have

$$\log\left|\frac{g_i(z)}{g_{n+1}(z)}\right| \leq \frac{1}{2\pi}\int_0^{2\pi}\log\left|\frac{g_i(Re^{i\theta})}{g_{n+1}(Re^{i\theta})}\right|d\theta + \frac{2}{\sigma-1}\frac{1}{2\pi}\int_0^{2\pi}\left|\log\left|\frac{g_i(Re^{i\theta})}{g_{n+1}(Re^{i\theta})}\right|\right|d\theta$$

$$- \sum_{|\beta_\mu^{(i)}| < R} \log|g(z, \beta_\mu^{(i)})|, \quad (i = 1, 2, \ldots, n)$$

where

$$g(z, b) = \frac{R(z - b)}{R^2 - \bar{b}z}.$$

By Jensen formula

$$\frac{1}{2\pi}\int_0^{2\pi}\log\left|\frac{g_i(Re^{i\theta})}{g_{n+1}(Re^{i\theta})}\right|d\theta = N\left(R, \frac{g_{n+1}}{g_i}\right) - N\left(R, \frac{g_i}{g_{n+1}}\right) + K$$

$$N\left(R, \frac{1}{g_i}\right) - N\left(R, \frac{1}{g_{n+1}}\right) + K.$$

Using the first fundamental theorem we have

$$\frac{1}{2\pi}\int_0^{2\pi}\left|\log\left|\frac{g_i(Re^{i\theta})}{g_{n+1}(Re^{i\theta})}\right|\right|d\theta = m\left(R, \frac{g_i}{g_{n+1}}\right) + m\left(R, \frac{g_{n+1}}{g_i}\right) \leq 2T\left(R, \frac{g_i}{g_{n+1}}\right) + K$$

In addition we have

$$\sum_{|\beta_\mu^{(i)}| < R} \log|g(z, \beta_\mu^{(i)})| \geq \sum_{|\alpha_\lambda| < R} \log|g(z, \alpha_\lambda)|.$$

Hence

$$\log\left|\frac{g_i(z)}{g_{n+1}(z)}\right| \leq N\left(R, \frac{1}{g_i}\right) - N\left(R, \frac{1}{g_{n+1}}\right) + \frac{4}{\sigma - 1}T\left(R, \frac{g_i}{g_{n+1}}\right)$$

$$- \sum_{|\alpha_\lambda| < R} \log|g(z, \alpha_\lambda)| + K. \quad (i = 1, 2, \ldots, n) \tag{3}$$

Obviously, (3) is also valid for i = n + 1.

We write again

$$g(z) = \max_{1 \leq i \leq n+1} |g_i(z)|.$$

$$\mu(r, \, g) = \frac{1}{2\pi n} \int_o^{2\pi} \log g(re^{i\theta}) d\theta,$$

thus

$$|\mu(r, \, g) - \mu(r, \, A)| \leq K. \tag{4}$$

By lemma 1 in (4) we have

$$T(R, \, \frac{g_i}{g_{n+1}}) < n\mu(R, \, g) + K, \quad (i = 1, \, 2, \, \ldots, \, n + 1) \tag{5}$$

so that from (3), (4) and (5) we deduce

$$\log g(z) - \log|g_{n+1}(z)| \leq \max_{1 \leq i \leq n+1} N(R, \, \frac{1}{g_i}) - N(R, \, \frac{1}{g_{n+1}}) +$$

$$\frac{4n}{\sigma-1} \mu(R, \, A) - \sum_{|\alpha_\lambda|<R} \log|g(z, \, \alpha_\lambda)| + K$$

In the above inequality we set that $z = re^{i\theta}$ and integrate with respect to $\theta$, we obtain

$$n\mu(r, \, A) - \frac{1}{2\pi}\int_o^{2\pi} \log|g_{n+1}(re^{i\theta})|d\theta \leq \max_{1 \leq i \leq n+1} N(R, \, \frac{1}{g_i}) - N(R, \, \frac{1}{g_{n+1}})+$$

$$\frac{4n}{\sigma-1} \mu(R, \, A) - \sum_{|\alpha_\lambda|<R} \frac{1}{2\pi}\int_o^{2\pi} \log|g(re^{i\theta}, \, \alpha_\lambda)|d\theta + K$$

Since

$$\frac{1}{2\pi}\int_o^{2\pi} \log|g_{n+1}(re^{i\theta})|d\theta = N(r, \, \frac{1}{g_{n+1}})+ K$$

and

$$\sum_{|\alpha_\lambda|<R} \frac{1}{2\pi}\int_o^{2\pi} \log|g(re^{i\theta}, \, \alpha_\lambda)|d\theta = N(r, \, \frac{1}{g_{n+1}}) - N(R, \, \frac{1}{g_{n+1}}),$$

we have

$$n\mu(r, \, A) \leq \max_{1 \leq i \leq n+1} N(R, \, \frac{1}{g_i}) + \frac{4n}{\sigma-1}\mu(R, \, A) + K \tag{6}$$

Let $\eta = 1 - \min\limits_{1 \leq i \leq n+1} \delta(a_i, \, w)$, where $\delta(a_i, \, w)$ (i = 1, $\ldots$, n+1) are the

deficiencies of w(z). We take C' and C such that

$$\eta < C' < C < 1$$

Y. GU

thus for sufficiently large r

$$\frac{1}{n} N(r, \frac{1}{g_i}) < C' \, T(r) \qquad (i = 1, 2, \ldots, n + 1) \tag{7}$$

In (6) we take that $\sigma = 1 + \frac{4}{C(1-C)}$. Using again (2) and (7) we obtain

$$T(r) < T(\sigma r) \left\{ C' + \frac{4}{\sigma-1} + \frac{K}{T(r)} \right\}.$$

Hence there exists a positive number $r_0$ such that for $r \geq r_0$

$$T(r) < T(\sigma r) \left\{ C + \frac{4}{\sigma-1} \right\} = C(2-C) \, T(\sigma r),$$

thus

$$\frac{T(\sigma^k r_0)}{T(\sigma^{k-1} r_0)} \; > \; \frac{1}{C(2-C)}. \qquad (k = 1, 2, \ldots)$$

Now we consider that $\sigma^m r_0 \leq r \leq \sigma^{m+1} r_0$. Thus

$$\frac{\log T(r)}{\log r} \; > \; \frac{\log T(\sigma^m r_0)}{\log (\sigma^{m+1} r_0)} \; > \; \frac{m\log \frac{1}{C(2-C)} + \log T(r_0)}{(m+1)\log\sigma + \log r_0} \; ,$$

so that

$$\mu = \frac{\lim}{r \to +\infty} \; \frac{\log T(r)}{\log r} \; \geq \; \frac{\log \frac{1}{C(2-C)}}{\log (\sigma)} \; = \; \frac{\log \frac{1}{C(2-C)}}{\log (1 + \frac{4}{C(1-C)})} \; .$$

The proof of this theorem is thus completed.

If in (8) we let $C \to \eta$, then we have the following:

Theorem 2. Let $w(z)$ be a n-valued transcendental algebroid function with $n + 1$ deficient values $a_i$ ($i = 1, \ldots, n + 1$) such that $\delta(a_i, w) = 1$ ($i = 1, \ldots, n + 1$). Then the lower order of $w(z)$ is at least equal to 1.

## REFERENCES


[1]   A. Edrei and W.H.J. Fuchs, Trans. Amer. Math. Soc. 93 (1959), 292-328.

[2]   M. Ozawa, Ko̊dai Math. Sem. Rep. 22 (1970), 122-127.

[3]   G. Valiron, Bull. Soc. Math. 59 (1931), 17-39.

[4]   N. Toda, Ko̊dai Math. Sem. Rep. 22 (1970), 114-121.



DEPARTMENT OF MATHEMATICS
NAN-CHUN NORMAL COLLEGE
SZE-CHWAN PROVINCE
PEOPLE'S REPUBLIC OF CHINA

Contemporary Mathematics
Volume 25, 1983

# ADMISSIBLE SOLUTIONS OF ORDINARY DIFFERENTIAL EQUATIONS

Yuzan He                    Xiuzhi Xiao

## I.  SUMMARY

J. Malmquist [12] has proved the following important theorem:  If the differential equation with rational coefficients

$$\frac{dw}{dz} = \frac{\sum_{i=o}^{p} a_i(z)w^i}{\sum_{j=o}^{q} b_j(z)w^j} \tag{1}$$

has a trascendental mermorphic solution, then (1) must be degenerate into a Riccati equation, i.e., $q=o$ and $p \leq 2$.

It is noticeable that the transcendental solutions among all solutions of the equation with rational coefficients play important role and bounds for p and q are precise.  In 1933, K. Yosida [16] presented a elegant proof and extensions of the Malmquist theorem by using the Nevanlinna's value distribution theory.  A systematic study of the implications of the Nevanlinna theory for ordinary differential equations has been undertaken by H. Wittich and his students, starting in 1950.  A series of further results were obtained by A.A. Gol'dberg, [15], I. Laine, [10], S. Bank, [1], E. Hille, [9], and C.C. Yang, [15].  In most circumstances the equations have solutions with finite branches, so the relative problem about algebroid solutions was considered in [4][12][17].  Recently F. Gackastatter and I. Laine considered the generalized algebraic differential equation

$$\Omega(z; w,w',\ldots w^{(n)}) = R(z,w) \tag{2}$$

where $\Omega(z, w,w',\ldots w^{(n)}) \equiv \Sigma a_{(i)}(z)(w)^{i_o}(w')^{i_1}\ldots(w^{(n)})^{i_n}$ is a differential polynomial with meromorphic coefficients $\{a_{(i)}(z)\}$ and $R(z,w) \equiv \dfrac{P(z,w)}{Q(z,w)} \equiv$

$\sum_{i=o}^{p} a_i(z)w^i \Big/ \sum_{j=o}^{q} b_j(z)w^j$ is an irreducible rational function in w with the

meromorphic coefficients.

They showed the rate of the mermorphic solution under consideration is larger compared to that of the coefficients, and gave a definition of admissible solution and obtained a theorem of the Malmquist type.

In this paper we shall discuss the same problem and give a definition of admissible and obtain a relative satisfactory result.  Here we use the standard notations and results of the Nevanlinna's theory, see, e.g., [6] [13].

Let
$$S_1(r,w) = \Sigma T(r,a_{(i)}) + \Sigma T(r,a_i) + \Sigma T(r,b_j) \tag{3}$$

If an algebroid solution $w(z)$ of (2) satisfies

$$\overline{\lim_{r\to\infty}} \ \frac{S_1(r,w)}{T(r,w)} = 0 \tag{4}$$

outside of a possible exceptional set E with a finite linear measure,
Gackstatter and Laine called $w(z)$ an admissible solution.  They proved the
following result:

Theorem A.  If the equation (2) admits at least an admissible algebroid
solution with $\nu$ branches, then

$$q \leq 4\overline{\mu}(\nu-1) \text{ and } p \leq q + \Delta \tag{5}$$

where $\overline{\mu}=\mathrm{Max}\{\overline{\mu}_1\}$, $\overline{\mu}_i=i_1 + 2i_2+\ldots+ni_n$ ;  $\Delta=\mathrm{Max}\{\Delta_i\}$ , $\Delta_i=i_0+2i_1+\ldots+(n+1)i_n$.

We give a definition of an admissible solution as follows

Let $w(z)$ be an algebroid function and $U(r)=r^{\rho(r)}$ be a type function of
$T(r,w)$, where $\rho(r)$ is a precise order of Valiron (or of Chung) if $U(r)$ is of
finite order (or infinite order).  Then $w(z)$ is an admissible function respect
to the coefficients of the equation (2), if

$$\overline{\lim_{r\to\infty}} \ \frac{S_1(r,w)}{U(r)} = 0 \tag{6}$$

outside of a possible exceptional set E with a finite linear measure.  The
solution of the equation (2) is admissible, if $w(z)$ satisfies (6).

Since $T(r,w)\leq U(r)$, therefore any admissible solution in the sense of (4)
must be an admissible solution in the sense of (6).

We have

Theorem 1.  If the equation (2) admits at least an admissible solution
with $\nu$ branches, then

$$q \leq 2\sigma(\nu-1) \quad \text{and } p \leq q + \lambda +\overline{\mu}\nu(1-\theta(w,\infty)) \tag{7}$$

where $\sigma=\mathrm{Max}\{\sigma_i\}$, $\sigma_i=i_1+3i_2+\ldots+(2n-1)i_n$; $\lambda=\mathrm{Max}\{\lambda_i\}$, $\lambda_i=i_0+i_1+\ldots+i_n$ ;

$$\theta(w,\infty)=1-\overline{\lim} \ \frac{N(r,w)}{T(r,w)}.$$

Because $\sigma<2\overline{\mu}$ always, the upper bound of $q$ in Theorem 1 is more precise
than one in Theorem A.

Finally, we give some example to show the bounds in our theorem can
be reached.

II.  Auxiliary results

In this section we shall give three lemmas.

Lemma 1.  Let

$$P(z,w) \equiv \sum_{i=o}^{p} a_i(z)w^i \ , \tag{8}$$

where $\{a_i(z)\}$ are mermorphic functions of z.  If w(z) is an algebroid function, then

$$T(r,p(z,w(z)))=pT(r,w) + 0\{\Sigma T(r,a_i)\} \ . \tag{9}$$

Proof.  It has been shown in [7] that

$$PN(r,w)-(p+i)\{\Sigma N(r,a_i)+ \Sigma N(r, \frac{1}{a_i})\} +0(1)$$

$$\leq N(r,p(z,w(z)))$$

$$\leq pN(r,w)+ N(r,a_i) +0(1)$$

or

$$N(r,p(z,w(z)))=pN(r,w)+0 \{\Sigma m(r,a_i)+\Sigma m(r, \frac{1}{a_i})\} \ . \tag{10}$$

We proceed to prove

$$m(r,p(z,w(z)))=pm(r,w)+0 \{\Sigma m(r,a_i)\div \Sigma m(r,\frac{1}{a_i})\} \ . \tag{11}$$

We denote by L a curve connecting all branch points of w(z) and $C'=C\backslash L$, then every branch $w_j(z)$ (j=1,2,...,$\nu$) of w(z) is single-valued in C'.  Set

$$P_j(z) \equiv P(z,w_j(z)) = \Sigma a_i(z)(w_j(z))^i$$

and write $E_r=\{z: |z|=r\}$ , $E_1^j =\{z\epsilon E_r, |W_j(z)|<1\}$  and $E_2^j =E_r\backslash E_1^j$ , so that

$$\frac{1}{2\pi} \int_{E_r} \ln^+|P_j(z)|d\theta =\frac{1}{2\pi} \int_{E_1^j} \ln^+|P_j(z)|d\theta +\frac{1}{2\pi} \int_{E_2^i} \ln^+|P_j(z)|d\theta, \quad z=re^{i\theta} .$$

It is easy to show

$$\frac{1}{2\pi} \int_{E_1^j} \ln^+|P_j(z)|d\theta \leq \frac{1}{2\pi} \int_{E_1^j} \Sigma\ln^+|a_i(z)|d\theta + \ln(p+1)$$

noting that

$$|P_j(z)| \leq |w_j(z)|^p \sum_{i=o}^{p} |a_i(z)| \quad \text{on } E_2^j, \text{ we have}$$

$$\frac{1}{2\pi} \int_{E_1^j} \ln^+|P_j(z)|d\theta \leq \frac{p}{2\pi} \int_{E_2^j} \ln^+|w_j(z)|d\theta + \sum_{i=o}^{p} \frac{1}{2\pi} \int_{E_2^j} \ln^+|a_i(z)|d\theta + \ln(p+1) .$$

Combining the above two expressions we get

$$m(r,P(z,w)) =\frac{1}{2\pi\nu} \sum_{j=1}^{\nu} \int_{E_r} \ln^+|P_j(z)|d\theta$$

$$\le pm(r,w) + \sum_{i=0}^{p} m(r,a_i) + O(1) \; . \tag{12}$$

On the other hand, we modify $P_i(z)$ in the form

$$P_j(t) = a_p(z) \sum_{i=0}^{p} A_i(z)(w_j(z))^{p-i} \; ,$$

where $A_i(z) = \dfrac{a_{p-i}(z)}{a_p(z)}$ . $i=0,1,\ldots,p$. For each $z\varepsilon C$ we define

$$A(z)=\underset{1\le i\le p}{\mathrm{Max}}\, \{1, |A_i(z)|^{\frac{1}{i}}\}$$

Let $\tilde{E}_1^j = \{z\varepsilon E_r, |w_j(z)| < 2A(z)\}$ , $\tilde{E}_2^j = E_r/\tilde{E}_1^j$ . Hence in $\tilde{E}_2^j$

$$|P_j(z)| \ge |a_p(z)||w_j(z)|^p\{1-\sum_{i-1}^{p}(\frac{A_i(z)}{|w_j(z)|})^i\} \ge a_p(z)||w_j(z)|^p/2^{p+1}$$

and hence

$$|w_j(z)|^p \le \begin{cases} (2A(z))^p, & \text{if } z\varepsilon\tilde{E}_1^j \\[2ex] \dfrac{2^{p+1}|P_j(z)|}{|a_p(z)|}, & \text{if } z\varepsilon\tilde{E}_2^j \end{cases}$$

Thus

$$\frac{p}{2\pi}\int_{E_r} \ln^+|w_j(z)|d\theta = \frac{1}{2\pi}\int_{E_1^j}\ln^+|W_j(z)|^p d\theta + \frac{1}{2\pi}\int_{E_2^j}\ln^+|w_j(z)|^p d\theta$$

$$\le \frac{1}{2\pi}\int_{\tilde{E}_1^j}\ln^+(2A(z))^p d\theta + \frac{1}{2\pi}\int_{\tilde{E}_1^j}\ln^+(\frac{2^{p+1}|P_j(z)|}{|a_p(z)|})\,d\theta$$

$$\le p\sum_{i=0}^{p-1} m(r,a_i)+p^2 m(r,\frac{1}{a_p})+ \frac{1}{2\pi}\int_{\tilde{E}_1^j}\ln^+|P_j(z)|d\theta + O(1)\; .$$

Therefore

$$pm(r,w) \le m(r,p(z,w)) + p\sum_{i=0}^{p-1} m(r,a_i)+p^2 m(r,\frac{1}{a_p}) + O(1)\; . \tag{13}$$

Combining (12) with (13) we deduce (11). Lemma 1 follows by addition.

  Examining that proof of Lemma 1 presented in [7], we verify

  Lemma 2. Let $Q(z,w) = \sum_{j=0}^{q} b_j(z)w^j$, $\Omega(z,w) \equiv \sum_{(i)} a_{(i)}(z)(w)^{i_0}\ldots(w^{(i)})^{i_n}$

and $\Omega_1(z,w) \equiv \Sigma\ b_{(j)}(z)(w)^{j_o} \dots (w^{(m)})^{j_m}$.  If $w(z)$ is an algebroid function with $\nu$ branches and satisfies

$$Q(zw(z))\Omega(z,w(z)) = \Omega_1(z,w(z))$$

then we have

$$m(r,\Omega) = O\{\Sigma T(r,a_{(i)}) + \Sigma T(r,a_i) + \Sigma T(r,b_j)\} + S(r,w) \qquad (14)$$

for $q \geq \lambda'$, where $\lambda'=\mathrm{Max}\{\lambda'_j\}$, $\lambda'_j = j_o + j_1 + \dots + j_m$; $S(r,w)=o\{\ln(rT(\rho,w))\}$.

We shall prove the following lemma which was given in other form [4].

Lemma 3.  Let $R(z,w) \equiv P(z,w)/Q(z,w) \equiv \sum_{i=o}^{p} a_i(z)w^i / \sum_{j=o}^{q} b_j(z)w^j$ be an irreducible rational function in $w$ with the meromorphic coefficients $\{a_i(z)\}$ and $\{b_j(z)\}$.  If $w(z)$ is an algebroid function then

$$T(r,R(z,w(z))) = sT(r,w) + O\{\Sigma T(r,a_i) + \Sigma T(r,b_j)\} \qquad (15)$$

where $s = \mathrm{Max}\{p,q\}$.

Proof.  Without loss of generality we can assume $\deg P(z,w)=p \geq \deg Q(z,w) = q$, for otherwise we may consider $\dfrac{1}{R(z,w)}$ instead of $R(z,w)$.  By using the division algorithm we have

$$P(z,w)=S_1(z,w)Q(z,w)+T_1(z,w), \quad \deg S_1 = p-q, \deg T_1 = t_1 < \deg Q$$

$$Q(z,w) = S_2(z,w)T_1(z,w)+T_2(z,w), \quad \det T_2 = t_2 < t_1$$

$$\dots\dots$$

$$T_{m-2}(z,w)=S_m(z,w)T_{m-1}(z,w)+T_m(z), \quad \deg T_m = t_m = o .$$

Putting these expressions together we get

$$T_m(z)=P(z,w)S(z,w)+Q(z,w)T(z,w) \qquad (16)$$

Since $P(z,w)$ is prime to $Q(z,w)$, $T_m(z) \not\equiv 0$.  Dividing both side of (16) by $T_m(z)$ we have $p(z,w)U(z,w)+Q(z,w)V(z,w)=1$ where $\deg U(z,w) \leq q-1$, $\deg V(z,w) \leq p-1$.

It is clear that the coefficients of $U(z,w)$ and $V(z,w)$ are the rational functions of $\{a_i(z)\}$ and $\{b_j(z)\}$, and that $U(z,w)$ is prime to $V(z,w)$ in the field of meromorphic functions.  Since $\deg U(z,w)+p = \deg V(z,w)+q$ and $p \geq q$ we have $\deg U(z,w) \leq \deg V(z,w)$.

By means of Lemma 1 and Nevanlinna's first fundamental theorem time after time, it follows

$$T(r,R(z,w)) = T\left(r,\frac{P(z,w)}{Q(z,w)}\right) \leq T(r,S_1(z,w))+T\left(r,\frac{T_1(z,w)}{Q(z,w)}\right)+O(1)$$

$$=T(r,S_1(z,w)) + T(r,\frac{(Q(z,w)}{T_1(z,w)}) + O(1)$$

$$\leq T(r,S_1(z,w))+\ldots+T(r,S_m(z,w))+T(r,\frac{T_{m-1}}{T_m}) + O(1)$$

$$=(p-q)T(r,w)+(q-t_1)T(r,w)+\ldots+(t_{m-1}-t_m)T(r,w)$$

$$+O\{\Sigma T(r,a_i) + \Sigma T(r,b_j)\}$$

$$=pT(r,w)+ O\{\Sigma T(r,a_i)+ \Sigma T(r,b_j)\} \ .$$

Thus

$$T(r,R(z,w)) \leq pT(r,w) + O\{\Sigma T(r,a_i) + \Sigma T(r,b_j)\}. \tag{17}$$

On the other hand, similar to the proof in [4] we get

$$T(r,R(z,w)) \leq pT(r,w) - O\{\Sigma T(r,a_i) + \Sigma T(r,b_j)\} \ . \tag{18}$$

Combining (17) with (18) we deduce (15).

III.   The proof of Theorem 1.

First we have

$$\Sigma a_{(i)}(z)(w)^{i_o}\ldots(w^{(n)})^{i_n} = P_1(z,w) + \frac{P_2(z,w)}{Q(z,w)} \ ,$$

then we get the modified equation

$$Q(z,w)F(z,w) = P_2(z,w) \ , \tag{2'}$$

where

$$F(z,w) \equiv \Sigma a_{(i)}(z)(w)^{i_o}\ldots(w^{(n)})^{i_n} - P_1(z,w)$$

and $\deg P_1(z,w) = p-q$, $\deg P_2(z,w)<q$. It is obvious that every admissible solution of (2) is also an admissible solution of (2'). According to Lemma 2 we have

$$m(r,F(z,w)) = O\{\Sigma T(r,a_{(i)}) + \Sigma T(r,a_i) + \Sigma T(r,b_j)\} + S(r,w) \tag{19}$$

where $S(r,w) = O\{\ln(rT(\rho,w))\}$ ($\rho$ to be defined).

We proceed to prove the following inequality:

$$N(r,F(z,w))\leq 2\sigma(\nu-1)T(r,w) + O\{\Sigma T(r,a_{(i)}) + \Sigma T(r,a_i) + \Sigma T(r,b_j)\}. \tag{20}$$

To prove this we need some estimates on the number of the poles of $F(z,w)$. In fact, the ploes of $F(z,w(z))$ may rise from one of the following three cases: (i) the poles of $\{a_{(i)}(z)\}$ and of the coefficients of $P_1(z,w(z))$; (ii) the poles of $w(z)$; (iii) the poles of some $w^{(j)}(z)$ $(j=1,2,\ldots,n)$ which is not the pole of $w(z)$.

Since the coefficients of $P(z,w)$ are the rational functions of $\{a_i(z)\}$ and $\{b_j(z)\}$, the poles which rise from case (i) contribute $O\{\Sigma T(r,a_{(i)}) +$

$\Sigma T(r,a_i) + \Sigma T(r,a_i) + \Sigma T(r,b_j)\}$ to $N(r,F(z,w))$.

For the case (ii). We first write $F(z,w)$ into the form

$$F(z,w) = \frac{P_2(z,w)}{Q(z,w)} = \frac{\displaystyle\sum_{i=0}^{P_1} \tilde{A}_i(z)w^i}{\displaystyle\sum_{j=0}^{q} b_j(z)w^j} \quad ,$$

where $\{\tilde{A}_i(z)\}$ are the rational functions of $\{a_i(z)\}$ and $\{b_j(z)\}$. Let $z$ be a pole of the function $a(z)$, we denote by $\tau(z,a)$ its multiple order. It is obvious that

$$\tau(z,P_2) \le p_2\tau(z,w) + \sum_{i=0}^{P_2} \tau(z,\tilde{A}_i)$$

and

$$\tau(z,Q) \ge q\,\tau(z,w) - \sum_{j=0}^{q} \tau(z,b_j) \quad .$$

If $q\tau(z,w) - \Sigma\tau(z,b_j) \ge 0$, noting $q > p_1$ we have

$$\tau(z,F) \le \tau(z,P_2) - \tau(z,Q) \le \Sigma\tau(z,\tilde{A}_i) + \Sigma\tau(z,b_j) \tag{21}$$

If $q\tau(z,w) - \Sigma\tau(z,b_j) < 0$, then

$$\tau(z,w) < \frac{1}{q} \Sigma\tau(z,b_j) . \tag{22}$$

Suppose now that $\eta$ branches of $w(z)$ have a pole at $z$ so that $\Omega_{(i)}(z) \equiv (w(z))^{i_0}$ $\ldots(w^{(n)}(z))^{i_n}$ has a pole at $z$ with multiplicity $\tau(z,\Omega_{(i)}(z)) = i_0\tau(z,w) + i_1$ $(\tau(z,w)+\eta)+\ldots+i_n(\tau(z,w)+n\eta) = \lambda_i\tau + \eta\bar{\mu}_i$, where $\lambda_i = i_0 + i_1 + \ldots + i_n$, $\bar{\mu}_i = i_1 + 2i_2 + \ldots + ni_n$. Noting $\tau(z,P_1) \le p_1\tau(z,w)$ and (22) we have

$$\tau(z,F) \le (\lambda+p_1)\tau(z,w) + \bar{\mu}\nu \le (\lambda+p_1+\mu\nu)\tau(z,w)$$

$$\le \frac{\lambda+p_1+\bar{\mu}\nu}{q} \Sigma\tau(z,b_j) \tag{23}$$

where $\lambda = \mathrm{Max}\{\lambda_i\}$, $\bar{\mu} = \mathrm{Max}\{\bar{\mu}_i\}$.

(21) and (23) show that the multiple order of the poles of $F(z,w)$ which rise from the case (ii) can be bounded by that of the coefficients, namely, these poles contribute $O\{\Sigma\tau(r,a_i) + \Sigma\tau(r,b_j)\}$ to $N(r,F(z,w))$.

Case (iii). We suppose that $\eta$ branches of $w(z)$ take $\alpha \ne \infty$ at $z_0$ which is a pole of $w^{(j)}(z)$. Clearly we have

$$w^{(j)}(z) = b(z-z_0)^{(\tau(z_0,w=\alpha)-j\eta)/\eta}(1+\phi(z-z_0))$$

58            Y. HE                       X. XIAO

where $\phi(z-z_o)$ is equal to zero at $z_o$. Thus

$$\tau(z,w^{(j)})=j\eta - \tau(z_o,w=\infty) \le j\eta - 1 \le (2j-1)(\eta-1).$$

Therefore

$$\tau(z_o,\ \Omega_{(i)}) \le i_1(\eta-1)+3i_2(\eta-1)+\ldots+(2n-1)i_n(\eta-1) = \sigma_i(\eta-1),$$

hence

$$\tau(z_o,F) \le \sigma(\eta-1) \tag{24}$$

where $\sigma=\mathrm{Max}\{\sigma_i\}$. (24) shows that these poles of $F(z,w(z))$ contribute $_\Omega N_n(r,w)$ to $N(r,F(z,w))$ at most. On the other hand, we have [13]

$$N_n(R,w) \le 2(\nu-1)T(r,w)+O(1).$$

putting these results together we deduce (22).

Now we proceed to prove the first assertion of Theorem 1. By combining (19) with (20) we get

$$T(r,F) \le 2\sigma(\nu-1)T(r,w)+S(r,w)+S_1(r,w)$$

where $S(r,w)=O\{\ln(rT(\rho,w))\}$, $S_1(r,w)=O\{\Sigma T(r,a_{(i)})+ \Sigma T(r,a_i)+\Sigma T(r,b_j)\}$.

By Lemma 3

$$T(r,F(z,w))=T\left(r,\frac{P_2(z,w)}{Q(z,w)}\right) = qT(r,w)+O\{\Sigma T(r,a_i)+ \Sigma T(r,b_j)\}$$

thus

$$qT(r,w) \le 2\sigma(\nu-1)T(r,w)+S(r,w)+S_1(r,w). \tag{25}$$

Now set $\rho=r+w(U(r))$ and note that $T(\rho,w) \le U(p)$, $\lim\limits_{r\to\infty} \dfrac{U(\rho)}{U(r)} = 1.$

we have

$$S(r,w)=O\{\rho(r)\ln r\}.$$

Therefore (25) becomes

$$qT(r,w) \le 2\sigma(\nu-1)T(r,w)+O\{p(r)\ln r\} + S_1(r,w). \tag{26}$$

We choose a sequence $r_n\to\infty$ such that $T(r_n,w)=U(r_n)$. Dividing both side of (26) by $T(r,w)$ and noting the definition of admissible solution we get

$$q \le 2\sigma(\nu-1) + \lim\limits_{r\to\infty} \left\{\frac{O\{\rho(r)\ln r\}}{T(r,w)}+ \frac{S_1(r,w)}{T(r,w)}\right\}$$

$$\le 2\sigma(\nu-1) + \lim\limits_{r_n\to\infty} \left\{ \frac{O\{\rho(r_n)\ln r_n\}}{T(r_n,w)} + \frac{S_1(r_n,w)}{T(r_n,w)} \right\}$$

$$= 2(\nu-1) + \lim\limits_{r_n\to\infty} \left\{\frac{O\{\rho(r_n)nr_n\}}{U(r_n)} + \frac{S_1(r_n,w)}{U(r_n)} \right\}= 2\sigma(\nu-1)$$

This prove the first inequality of (7).

To prove the second assertion of Theorem 1 we first can write (2) as follows

$$P(z,w) = Q(z,w)\,\Omega(z;w,w', \ldots,w^{(n)})\ .$$

Since $P(z,w)$ is a polynomial in $w$ which doesn't contain any derivative $w^{(j)}(z)$, its poles can rise from the poles of $w(z)$ and that of the coefficients. It is also true that the poles of $Q(z,w)\,\Omega(z,w,\ldots,w^{(n)})$ can only rise from the poles of $w(z)$ and that of the coefficients. Thus we have the estimation

$$\tau(z,Q\Omega) \leq q\tau(z,w) + \lambda\tau(z,w) + \bar{\mu}\nu + \Sigma\tau(z,b_j) + \Sigma\tau(z,a_{(i)})\ .$$

Thus again

$$N(r,Q\Omega) \leq (q+\lambda)N(r,w)+\bar{\mu}\nu\bar{N}(r,w) + \Sigma N(r,b_j) + \Sigma N(r,a_{(i)})\ .$$

On the other hand, applying (12) to $Q(z,w)$ and noting the fundamental lemma on logarithmic derivative

$$m(r,Q\Omega) \leq m(r,Q) + m(r,\Omega(z;w,\tfrac{w'}{w}w,\ldots, \tfrac{w^{(n)}}{w}w))$$

$$\leq (q+\lambda)m(r,w) + S(r,w) + S_1(r,w).$$

In virtue of Lemma 1 and above two inequalities we get

$$pT(r,w) \leq (q+\lambda)T(r,w)+\bar{\mu}\nu\bar{N}(r,w) + S(r,w) + S_1(r,w).$$

By an argument similar to the proof for the first inequality of (7) the second assertion follows. This completes the proof of Theorem 1.

IV. Examples

The following two examples show that the bounds in Theorem 1 can be reached.

Example 1. It is easy to show that the algebroid function defined by

$$a(z)w^2+\tan z\ w + b(z)=o \tag{27}$$

is an admissible solution of the equation

$$\frac{dw}{dz} = -\frac{a^2(z)w^4+a'(z)w^3+(1+2a(z)b(z))w^2+b'(t)w+b^2(z)}{a(z)w^2-b(z)}\ , \tag{28}$$

where $a(z)$ and $b(z)$ are polynomial in $z$. Since $\nu=q=2$ and $\sigma=\bar{\mu}=1$, the upper bound of $q$ is $2\sigma(\nu-1)=2$. It shows that the bound can be reached.

Example 2. The algebroid function $w(z)$ defined by

$$\cos z w^{\nu} - c(z)\sin z=0 \tag{29}$$

is an admissible solution of the equation

$$\frac{dw}{dz} = \frac{w^{2\nu}-c'(z)w^{\nu}-c^2(z)}{\nu c(z)w^{\nu-1}}$$

where $c(z)$ is a polynomial of $z$. Since $\rho=2\nu, q=\nu+1, \lambda=\overline{\mu}=\sigma=1$ and $\Delta=2, \theta(w,\infty)=o$, the upper bound of $p$ in Theorem A is $q+\Delta=\nu+1$. It shows that the conclusion on the bound of $p$ in Theorem A is not true when $\nu>1$. While the upper bound of $p$ in Theorem 1 is $q+\lambda+\overline{\mu}\nu=2\nu$, it shows that the bound can not be reduced.

Remark. If the coefficients $a(z)$, $b(z)$ (or $c(z)$) are mermorphic functions with order less than 1, then the algebroid function defined by (26) (or by (28)) is an admissible solutions of (27) (or (29)) in the sence of (6), but it is not easy to judge whether it is admissible in the sense of (4) or not.

## References


1. S. Bank, Some results on analytic and mermorphic solutions of algebraic differential equations, Advances in Math. 15 (1975), 41-62.

2. S. Bank and I. Laine, On the growth of mermorphic solutions of linear and algebraic differential equations, Math. Scand. 40 (1977), 119-126.

3. Chung Chi-tai, Sur les fonctions-types, Scientia Sinica, 10 (1961) No. 2.

4. F. Gackstatter and I. Laine, Zur Theorie der gewöhnlichen Differential-gleichungen in Komplexen, Ann. Polon. Math. 38 (1980) 259-287.

5. A.A. Gol'dberg, On single-valued solutions of first order differential equations, Ukrain Mat. Zumal. 8:3 (1956), 254-261.

6. W.K. Hayman, Mermorphic functions, Oxford University Press, 1964.

7. He Yuzan, On algebroid solutions of ordinary differential equations, Acta Math. Sinica, 24 (1981), 464-471.

8. He Yuzan and Xiao Xiuzhi, On Malmquist's theorem of algebraic differential equations, Kexue Tongbao, 28 (1983), 165-169.

9. E. Hille, Finiteness of the order of meromorphic solutions of some non-linear ordinary differential equations, Proc. Roy. Soc. Edinburgh Sect. A72 (1974), 331-336.

10. I. Laine, On the behavior of the solutions of some first order differential equations, Ann. Acad. Sci. Fenn. Ser. AI 497 (1971).

11. I. Laine, Admissible solutions of some generalized algebraic differential equations, Publ. Univ. Joensuu Ser. B10 (1974).

12. J. Malmquist, Sur les fonctions à un nombre fini des branches définies par les équations differentielles du premier order, Acta Math. 36 (1913), 297-343.

13. G. Valiron, Sur la dérivée des fonctions algébroïdes, Bull. Soc. Math. France 59 (1931), 17-39.

14. H. Wittich, Neuere Untersuchungen über eindeutige analytische Funktionen, Springer-Verlag, 1955.

15. C.C. Yang, On meromorphic solutions of generalized algebraic differential equations, Ann. Mat. Pura Appl. 91 (1972), 41-54.

16. K. Yosida, A generalization of a Malmquist's theorem, Japan J. Math. 9 (1933), 253-256.

17.  K. Yosida, On algebroid-solutions of ordinary differential equations,
     Japan J. Math. 10 (1934), 199-208.

Institute of Mathematics            Department of Mathematics
     Academia Sinica                   Wahan University
         Beijing                            China
         China

Contemporary Mathematics
Volume 25, 1983

SOLUTIONS WITH ONLY REAL ZEROS OF SECOND ORDER DIFFERENTIAL

EQUATIONS AND ZEROS OF DERIVATIVES OF A MEROMORPHIC FUNCTION

by

Simon Hellerstein, Li-Chien Shen, and Jack Williamson

In the series of papers [1], [2], and [4], the authors, exploiting earlier work [6] of B. Ja. Levin and I.V. Ostrovskii, succeeded in characterizing all entire functions which have, along with their first two derivatives, only real zeros - thus verifying an old conjecture of Polya. The solution of the Polya problem has led us to the formulation of a new conjecture for meromorphic functions. We state it as

Conjecture 1. Let F be real, transcendental, meromorphic, not entire, with only real poles. If F, F', and F" have only real zeros, then

(1)     $F(z) = A \tan (az + b) + B$

or

(2)     $F(z) = A[\tan(az + b) - (az + b)]$

where $A \neq 0$, $a \neq 0$, b and B are real constants.

We say that a function is real if it assumed only real values (or $\infty$) on the real axis and that it is strictly non-real if it is not a constant multiple of a real function.

The analogous characterization problem when F is strictly non-real has been solved in [4]. In [3] and [7] this problem was settled for reciprocals (of constant multiples) of real entire functions with only real zeros.

We have recently proved Conjecture 1 with the additional hypothesis that F' have no zeros. We have shown

Theorem 1. Let F be a real meromorphic function with only real zeros and real poles (and at least one of each). If F' has no zeros and F" only real zeros, then F has one of the forms

(3)     $F(z) = A \tan(az + b) + B$

(4)     $F(z) = A\dfrac{az+b}{cz+d}$

(5)     $F(z) = A\dfrac{(az+b)^2-1}{(az+b)^2}$

where A, B, a, b, c and d are real constants, A, a, and c nonzero and where in (4), $ad - bc \neq 0$.

For functions of finite order with only simple poles we require no assumption on the zeros of F". We have

Theorem 2. Let F be meromorphic of finite order, not entire, with only real zeros and poles. Suppose, in addition, that all the poles of F are simple and

                    S. Hellerstein, L. Shen, and J. Williamson

that F' has no zeros.  Then, if F is real it is of the form (3) or (4).  If
F is strictly nonreal it is of the form

(6)        $A[\tan(az + b) + i]$,

where $A \neq 0$, $a \neq 0$, and b are constants with a and b real.

Our proof of Theorems 1 and 2, which is somewhat lengthy, will be presented elsewhere [5].

Our object here will be to show that Theorems 1 and 2 (which verify
Conjecture 1 for a limited class of functions), when combined with the results
in [1], [2], [3], [4], and [7], shed light on the question of the existence of
solutions with only real zeros of second order differential equations.

In order to motivate our question and to understand its connection to
Theorems 1 and 2 we begin with several elementary observations concerning
solutions of the equation

(7)        $w'' + H(z)w = 0$.

If H is constant then (7) has a pair of linearly independent ($\ell$.i.)
solutions which have only real zeros.  For what other entire functions H does
(7) have two such solutions?  If $w_1$ and $w_2$ are $\ell$.i. solutions of (7) with only
real zeros then $w_1$ and $w_2$ have no common zeros and all their zeros are simple.
Moreover, since (7) has no first order term the Wronskian of $w_1$ and $w_2$ is a
constant.  It follows that the meromorphic function

(8)        $F = \dfrac{w_2}{w_1^2}$

has only real, simple zeros and poles, (if any), and that

(9)        $F' = \dfrac{C}{w_1^2}$ , C constant,

has no zeros.

In case H is a non-constant polynomial every nonzero solution of (1) is
entire of order at least 3/2, and the order is an integral multiple of 1/2
[8, pp. 68-70].  From (9) it is clear that F', and hence F, is of the same
order as $w_2$.  Moreover, either $w_1$ or $w_2$ must have some zeros.  Otherwise, F
would be entire of finite order exceeding one, and F as well as F' would be
zero free, which is impossible.  Thus we may assume, without loss of generality,
that $w_1$ has at least one zero (in the contrary case, we interchange $w_1$ and $w_2$
in (8).)  Thus, from the standpoint of solutions with only real zeros to (7)
with H a polynomial, our original question may be formulated as follows.  Which
mermorphic functions of integral or half-integral order, exceeding one, have
only real, simple zeros and poles, and have a zero-free derivative?  Appealing
to Theorem 2, the answer is none.  We therefore have

<u>Theorem 3</u>.  Suppose that H is a polynomial and that (7) has a pair of $\ell$.i. solutions which have only real zeros.  Then H is constant.

We remark that the asymptotic behavior as $|z| \to \infty$ of solutions of (7) when H is a polynomial has received much attention.  In view of this work Theorem 3 is not surprising, in the sense that the asumptotic behavior strongly suggests that one of the two $\ell$.i. solutions has some non-real zeros.  However, to the best of our knowledge, the actual existence of these zeros is not a direct sequence of known results.

We also note that the conclusion of Theorem 3 remains valid for $\ell$.i. solutions of the apparently more general equation $y'' + py' + q = 0$ where p and q are polynomials provided that $2p' + p^2 - 4q$ is not constant.  To see this we need only observe that if the latter condition is met, the change of variable $y = w \exp(-\frac{1}{2} \int P)$ transforms the given equation into one of the form (7) with H a non-constant polynomial.

We cannot hope, in Theorem 3, to eliminate the requirement that H be a polynomial.  Indeed, if we select an arbitrary non-constant entire function h and let $H = h'' - h'^2 - \exp(4h)$, then $w = \exp[-h \pm \int \exp(2h)]$ gives two $\ell$.i. solutions of (7) without any zeros.

In the case of transcendental H we can, however, assert the following:

<u>Theorem 4</u>.  Suppose that H is a transcendental entire function and that $w_1$ and $w_2$ are linearly independent solutions of (7).  If $w_1$, $w_1'$, and $w_2$ have only real zeros then

$$(10) \qquad H(z) = \frac{1}{2} c^2 e^{i(cz+d)} \cos(cz+d)$$

$$(11) \qquad w_1(z) = A \exp\{-\frac{1}{2}[i(cz+d) + e^{i(cz+d)}]\}$$

and

$$(12) \qquad w_2 = A \exp\{-\frac{1}{2}[i(cz+d) - e^{i(cz+d)}]\},$$

$A \neq 0$, $c \neq 0$, and d are constant, with c and d real.

Thus, except for the special choice (10) for H, equation (7) cannot have $\ell$.i. solutions which, along with the derivative of one of these solutions, have only real zeros.

To prove Theorem 4 we again introduce the auxiliary function F of (8).

Arguing as in the proof of Theorem 3 we conclude from the hypothesis of Theorem 4 that F is meromorphic of infinite order with all its zeros and poles simple, that F' has no zeros, and that F'' has only real zeros.  The proof is completed by noting that in view of Theorem 4 and the earlier results in [1], [2], [3], [4], and [7] the only mermorphic functions F with these properties are of the form $F(z) = A \exp(e^{i(cz+d)})$, where $A \neq 0$, $c \neq 0$, and d are constant with c and d real.  Then $w_1$ is found from (9), followed by H from (7) and $w_2$ from (8).  Details of this argument are given in [5].

Let us suppose now that in (7) H is a real entire function with only real zeros and that $w = w(z)$ is a solution of (7) which is real entire and has only real zeros.  It is, however, a conjecture (ca. 1915) of A. Wiman that the only such entire functions are those in the Laguerre-Polya class, i.e., of the form $e^{-az^2} g(z)$, where $a \geq 0$, $g = g(re^{i\theta})$ is real entire with only real zeros, and $\varlimsup_{r \to \infty} \{\log M(r,g) \big| r^2\} = 0$.  Since not all functions of Laguerre-Polya type can be solutions of (7), an analysis of the possibilities leads us to the following complementary conjecture.

<u>Conjecture 2</u>.  Suppose that H is real entire and has only real zeros and that (7) has a non-zero real entire solution with only real zeros.  Then H has one of the following forms;

$$(13) \qquad H(z) = az + b$$

$$(14) \qquad H(z) = \frac{a^2}{4} [2(2n+1) - (az+b)^2],$$

where a and b are real constants and n is a non-negative integer.

To prove Conjecture 2 it would be sufficient to assume that $\varlimsup_{r \to \infty} \{\log M(r,H) \big| r \log r\} < \infty$ — in particular that H is of order at most one.  This follows from the fact that in the contrary case every solution w of (7) satisfies $\varlimsup_{r \to \infty} \{\log \log M(r,w) \big| r \log r\} = \infty$, and a result of Levin and Ostrovskii [6] implies that for such w either w or w" has infinitely many non-real zeros.

We remark, finally, that if we add to the hypothesis of Conjecture 2 the additional requirement that the derivative of the solution also have only real solution of the Polya problem in [1] and [2].

REFERENCES

1.  S. Hellerstein and J. Williamson, <u>Derivatives of Entire Functions and a Question of Polya</u>, Trans. Amer. Math. Soc. 227 (1977), 227-249.

2.  _______________________________, <u>Derivatives of Entire Functions and a Question of Polya II</u>, Trans. Amer. Math. Soc. 234 (1977), 497-503.

3.  _______________________________, <u>The Zeros of the Second Derivative of the Reciporocal of an Entire Function</u>, Trans. Amer. Math. Soc. 263 (1981) 501-513.

4.  S. Hellerstein, L.C. Shen, and J. Williamson, Reality of the Zeros of an
    Entire Function and its Derivatives, Trans. Amer. Math. Soc., 275 (1983),
    319-331.

5.  ________________________________________, Real Zeros of Meromorphic
    Functions and Solutions of Second Order Differential Equations, (to appear).

6.  B. Ja. Levin and I.V. Ostrovskii, The Dependence of the Growth of an Entire
    Function on the Distribution of the Zeros of its Derivatives, Amer. Math.
    Soc. Translations (2) 32 (1963), 322-357.

7.  J. Rossi, The Reciprocal of an Entire Function of Infinite Order and the
    Distribution of the Zeros of its Derivative, Trans. Amer. Math. Soc. 270
    (1982), 667-683.

8.  H. Wittich, Neure Untersuchungen uber eindeutige analytische Funktionen,
    2nd Edition, Springer-Verlag (1968).

Department of Mathematics
University of Wisconsin
Madison, Wisconsin 53706

Department of Mathematics
The California Institute of Technology
Pasadena, California 91125

Department of Mathematics
University of Hawaii
Honolulu, Hawaii 96822

Contemporary Mathematics
Volume **25**, 1983

# A NECESSARY CONDITION FOR THE BLOCH ORDERING

by

J.S. Hwang and Peter Lappan

1. <u>Introduction</u>.  Let $D = \{z : |z| < 1\}$ denote the unit disc in the complex plane, and let E and E' be two subsets of the complex plane C.  We define the notation $E' \underset{B}{\leq} E$ to mean that for each function f analytic in D the condition

$$\sup \{1 - |z|^2)|f'(z)| : f(z) \in E\} < \infty$$

implies the condition

$$\sup \{1 - |z|^2)|f'(z)| : f(z) \in E'\} < \infty .$$

The ordering "$\underset{B}{\leq}$" defines a partial ordering of the subsets of the complex plane.  This ordering is called the <u>Bloch ordering</u>, and has been studied in [1] and [2].

For a set $E \subset C$ and a positive number K, let $N(E,K) = \{w \in C:$ there exists a point $t \in E$ with $|w - t| < K\}$.  We say that the set E is triply bunched if there exist a pair of functions $p_1$ and $p_2$ from E into itself and a pair of positive numbers $\alpha$ and $\beta$ such that for each $w \in E$ the three inequalities

$$\alpha < |w - p_1(w)| < \beta ,$$
$$\alpha < |w - p_2(w)| < \beta$$

and

$$\alpha < |p_1(w) - p_2(w)| < \beta$$

are satisfied.

In [1], the second author proved the following sufficient condition for the Bloch ordering:  if E is a triply bunched subset of the complex plane and if $K > 0$, then $N(E,K) \underset{B}{\leq} E$.  In the present paper, we prove that this sufficient condition is also necessary.

<u>Theorem</u>.  Let E be a subset of the complex plane and let $K > 0$.  In order that $N(E,K) \underset{B}{\leq} E$ it is both necessary and sufficient that E be triply bunched.

In view of the result cited above, the theorem will be proved when the necessity is established.  This will be done if we can show that, if E is not triply bunched, then there exists a function f analytic in D such that

$$\sup\{(1 - |z|^2)|f'(z)| : f(z) \in E\} < \infty$$

but

$$\sup\{(1 - |z|^2)|f'(z)| : f(z) \in N(E,K)\} = \infty.$$

We will demonstrate the existence of such a function by the construction of an appropriate Riemann surface.  In section 2, we will prove a lemma which may be

of some interest independent of the Theorem.  In section 3, we present a proof
of the Theorem.

2.  A Basic Lemma.  Let $f$ be a function analytic in $D$.  We can write
$f = \pi \circ f^*$, where $f^*$ is a conformal mapping from $D$ onto the simply connected
Riemann surface $R^*$ associated with the function $f$, and $\pi$ is the projection
mapping from $R^*$ onto $f(D)$.  If $z^*$ and $t^*$ are two points of $R^*$, the distance
between $z^*$ and $t^*$ on $R^*$ is defined to be the infimum of the lengths of all
curves $\pi(\gamma^*)$, where $\gamma^*$ is a curve in $R^*$ joining $z^*$ and $t^*$.  For $z \in D$, let
$d_1(z,f)$ denote the distance on $R^*$ between $f^*(z)$ and the nearest branch point
of $R^*$ (that is, a point $f^*(t)$ for which $f'(t) = 0$).  Let $d_2(z,f)$ denote the
Euclidean distance from $f(z)$ to the boundary of $f(D)$.  Finally, let $d(z,f)$ be
the distance on $R^*$ from $f^*(z)$ to the nearest branch point or the nearest
boundary point of $R^*$, whichever is smaller.  It is a well-known consequence of
Schwarz's Lemma that $d(z,f) \leq (1 - |z|^2)|f'(z)|$.  Also if $f(D)$ is a simply
connected region, then it is an easy consequence of the One Quarter Theorem
that $(1 - |z|^2)|f'(z)| \leq 4d_2(z,f)$ (see [3]).  We now give an improvement for
this last inequality.

Lemma.  Let $f$ be a function analytic in $D$, and let $0 \in f(D)$ be such that
$z \in f^{-1}(0)$ implies that $f^*(z)$ is a branch point of odd order.  Suppose $g(z) = \sqrt{f(z)}$ is such that $g(D)$ is simply connected.  If $z_0 \in D$ satisfies $d_1(z_0,f) = |f(z_0)| \leq d_2(z_0,f)$, then

$$(1 - |z_0|^2)|f'(z_0)| \leq 8(1 + \sqrt{2})\,\sqrt{d_1(z_0,f) \cdot d_2(z_0,f)} \quad .$$

Proof.  Let $G(t) = g((t+z_0)/(1+z_0 t))$ for $t \in D$.  Then $d_2(0,G) = d_2(z_0,g)$.
Let $z_1 \in f^{-1}(0)$ be such that $f^*(z_1)$ is the nearest branch point to $f^*(z_0)$.
Then $d_2(z_1,g) \leq \sqrt{d_1(z_0,f) + d_2(z_0,f)}$.  Thus, we have

$$d_2(z_0,g) \leq d_1(z_0,g) + \sqrt{d_1(z_0,f) + d_2(z_0,f)}$$
$$\leq \sqrt{d_1(z_0,f)} + \sqrt{d_1(z_0,f) + d_2(z_0,f)}$$
$$\leq (1 + \sqrt{2})\,\sqrt{d_2(z_0,f)} \quad .$$

By straightforward calculation,
$$|G'(0)| = (1 - |z_0|^2)|f'(z_0)| / |2G(0)| \quad .$$
Since $g(D)$ is simply connected, we have $|G'(0)| \leq 4d_2(0,G)$
and thus

$$(1 - |z_0|^2)|f'(z_0)| \leq 8|G(0)|d_2(0,G)$$
$$= 8\sqrt{d_1(z_0,f)}\,d_2(z_0,g)$$
$$\leq 8(1 + \sqrt{2})\,\sqrt{d_1(z_0,f)d_2(z_0,f)} \quad .$$

This form of the Lemma is adequate for our purposes, but it seems reasonable that some of the hypotheses can be weakened.  We would hope that the following statement is true:  There exists a continuous function H(x,y) defined on $[0,\infty] \times (0,\infty)$ such that $H(0,y) = 0)$ for each y and, if f is a function analytic in D then

$$(1 - |z|^2)\,|f'(z)| \leq H(d_1(z,f),\, d_2(z,f))$$

for each $z \in D$.  However, we are unable to prove this general a statement.

3.  <u>Proof of the Theorem</u>.  Suppose that E is not a triply bunched set. Then one of the following two possibilities must occur:  (1) There exists a sequence $\{w_n\}$ of points in E and a pair of sequences $\{r_n\}$ and $\{s_n\}$ of positive numbers such that $r_n \to 0$, $s_n \to \infty$, and $E \quad \{w : r_n < |w - w_n| < s_n\} = \emptyset$ for each positive integer n; or (2) there exists a pair of positive numbers $M_1$ and $M_2$, a pair of sequences $\{w_n\}$ and $\{w_n'\}$ of points in E, and a pair of sequences $\{r_n\}$ and $\{s_n\}$ of positive numbers such that $r_n \to 0$, $s_n \to \infty$, $M_1 \leq |w_n - w_n'| \leq M_2$, and

$$E \cap \{w : r_n < |w - w_n| < s_n\} \subset \{w : |w - w_n'| < r_n\}$$

for each positive integer n.  In each case, let $t_n = \min\{s_n/2,\ 1/r_n,\ |w_n|\}$.

In case (1), let $q_n = w_n$, and in case (2) let $q_n = (w_n + w_n')/2$.  In each case let $Q_n = \{w : |w - q_n| < t_n\}$. We may assume that $Q_n \cap Q_j = \emptyset$ for $j \neq n$, and, in addition, that $2r_n < t_n - M_2$ for each n, since we can always choose subsequences of points from E if necessary, and we can choose the sequence $\{s_n\}$ as slowly growing as we please.  For each positive integer n, let $L_n$ be a narrow channel joining $Q_n$ to $Q_{n+1}$ such that $Q = \bigcup_n (Q_n \cup L_n)$ is a simply connected region and such that if $w \in \bigcup_n L_n$, then the distance from w to the boundary of Q is at most 1.

In case (1), we attach a copy of $Q_n$, say $Q_n'$, to $Q_n$ by means of a single branch point at the center $q_n$, so that the resulting subsurface $Q_n \cup Q_n'$ is a copy of the Riemann surface for the function $z^2$ in an appropriate disc.  Let Q* be the Riemann surface which results from doing this to each $Q_n$.  Then Q* is a simply connected Riemann surface, so there exists a one-to-one function f* from D onto Q* such that $f = \pi \circ f*$ is analytic in D, where $\pi$ is the natural projection mapping from Q* to Q.  (Here $\pi$ is just the identity map in Q and for $w \in Q_n'$ we have that $\pi(w)$ is just the copied point on $Q_n$.)  The construction gives that $f(D) = Q$, and $f'(z) = 0$ if and only if $f(z) \in \{q_n : n \geq 1\}$.

In case (2), we modify Q by the following construction.  For each positive integer n, take countably infinitely many copies of $Q_n$ and attach each of these copies (including the original $Q_n$) to a pair of other "copies" in such a way that their union has a branch point of order one at each copy of $w_n$ and at each copy of $w_n'$, but no other branch points.  For a fixed n, the union of all the copies of $Q_n$ forms a Riemann surface which is conformally equivalent to the Riemann surface for the function $\sin z$ over an appropriate open set containing the real axis. Let $Q^*$ be the Riemann surface resulting from the total construction over all the $Q_n$'s. Then $Q^*$ is a simply connected Riemann surface, so there exists a one-to-one function $f^*$ from $D$ onto $Q^*$ such that $f = \pi \circ f^*$ is analytic in $D$, where again $\pi$ is the natural projection mapping from $Q^*$ to $Q$. (Again, if $w$ is in a copy of $Q_n$ then $\pi(w)$ is just the point of $Q_n$ for which $w$ is a copy.) As before, we have that $f(D) = Q$, and $f'(z) = 0$ if any only if $f(z) \in \bigcup_n \{w_n\} \cup \{w'_n\}$.

If either case, if $f(z) \in E$ then either $f(z) \in \bigcup_n L_n$ or $f(z) \in \bigcup Q_n$. If $f(z) \in \bigcup L_n$, then $(1 - |z|^2)|f'(z)| \leq 4d_2(z,f) \leq 4$. If $f(z) \in Q_n$ for some $n$, then $d_1(z,f) < r_n < t_n - M_2 - r_n \leq d_2(z,f)$, so we may apply the Lemma to the function $h(t) = f(t) - f(z)$ to obtain

$$(1 - |z|^2)|f'(z)| = (1 - |z|^2)|h'(z)| \leq 8(1 + \sqrt{2})\sqrt{d_1(z,f)d_2(z,f)}\ .$$

But $d_1(z,f) < r_n$ and $d_2(z,f) \leq 1/r_n$ by our construction, so we have

$$(1 - |z|^2)|f'(z)| \leq 8(1 + \sqrt{2}) \text{ whenever } f(z) \in E.$$

Let $K$ be a positive number. (In case (2), let $K < M_1/2$.) Let $\{z_n\}$ be a sequence of points in $D$ such that $K/4 \leq |f(z_n) - w_n| < K$ for each $n$. In case (1), let $g_n(t) = \sqrt{f((t+z_n)/1 + \overline{z}_n t)) - w_n}$. Then $g_n$ is analytic in $D$ and $d_2(0,g_n) \geq \sqrt{t_n} - 2\sqrt{K}$. A simple calculation gives that

$$|g'_n(0)| = (1 - |z_n|^2)|f'(z_n)|/(2|g_n(0)|) \geq d_2(0,g_n)$$

and thus

$$(1 - |z_n|^2)|f'(z_n)| \geq 2\sqrt{K/4}\ (\sqrt{t_n} - 2\sqrt{K}).$$

Thus, $(1 - |z_n|^2)|f'(z_n)| \to \infty$ as $n \to \infty$ and hence, in case (1), we have

$$\sup\{(1 - |z|^2)|f'(z)| : f(z) \in N(E,K)\} = \infty\ .$$

In case (2), the situation is more involved. Let $\Delta_n = \{z \in D : f(z) \in Q_n\} = f^{-1}(Q_n)$. Since $\pi^{-1}(Q_n)$ is a simply connected subset of $Q^*$, we have that $\Delta_n$ is simply connected also. Let $g_n$ be a conformal mapping from $\Delta_n$ onto the unit disc $D$ such that $g_n(z_n) = 0$. Recall that $q_n = (w_n + w'_n)/2$, and set $h_n(z) = ((w_n - w'_n)/2) \sin z + q_n$. Let $H_n$ be a one-to-one function on $\pi^{-1}(Q_n)$ such that $H_n(\pi^{-1}(Q_n))$ contains the real axis and $h_n \circ H_n \circ \pi^{-1}$ is the identity mapping on $Q_n$. Let $j_n = H_n \circ \pi^{-1}$, and let $G_n = j_n(Q_n)$. Then $G_n$ is a strip-like domain containing the real axis and such that $z \in G_n$ if and only if $(z + 2\pi) \in G_n$. Also, if $y_0$ is a fixed real number, let $d(n,x,y_0)$ denote the distance from the point $x + iy_0$ to the boundary of $G_n$. Then $d(n,y_0) = \inf\{d(n,x,y_0) : x \in R\} \to \infty$ as $n \to \infty$, since $t_n \to \infty$. Let $k_n$ be the conformal mapping from $D$ onto $G_n$ such that $k_n(0) = H_n(f^*(z_n))$. We now have the following diagram:

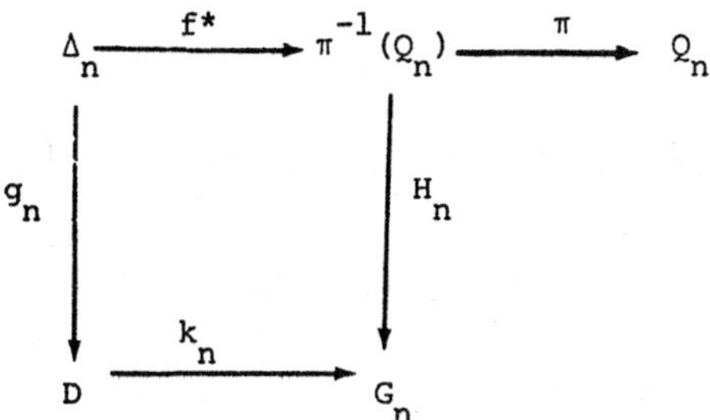

where $H_n(f^*(z_n)) = k_n(g_n(z_n))$. Hence, the mapping $g_n \circ (f^*)^{-1} \circ (H_n)^{-1} \circ k_n$ is a conformal mapping of D onto itself sending O to O, and thus is a rotation of D. Thus, there exists a number $\theta$, $0 \le \theta < 2\pi$, such that

$$(g_n \circ (f^*)^{-1} \circ (H_n)^{-1} \circ k_n)(t) = e^{i\theta} t$$

for each $t \in D$. Thus

$$((H_n)^{-1} \circ k_n)(t) = f^*(g_n^{-1}(e^{i\theta} t))$$

for each $t \in D$. By Schwarz's Lemma, since $g_n^{-1}$ maps D onto $\Delta_n \subset D$ we have

$$|(g_n^{-1})'(t)| \le (1 - |g_n^{-1}(t)|^2)/(1 - |t|^2)$$

for each $t \in D$. Setting $t = O$ and using that $g_n^{-1}(O) = z_n$, we have

$$|(g_n^{-1})'(O)| \le 1 - |z_n|^2 ,$$

and hence

$$|(f \circ g_n^{-1})'(O)| \le |f'(z_n)|(1 - |z_n|^2).$$

However, recalling that $j_n = H_n \circ \pi^{-1}$ and thus $\pi \circ H_n^{-1} = j_n^{-1}$, we have that

$$(j_n^{-1} \circ k_n)(t) = (f \circ g_n^{-1})(te^{i\theta})$$

for each $t \in D$, and thus

$$|(j_n^{-1} \circ k_n)'(O)| = |(j_n^{-1})'(k_n(O))||k_n'(O)| \le (1 - |z_n|^2)|f'(z_n)|.$$

Since

$$(j_n^{-1})'(z) = h_n'(z) = ((w_n - w_n')/2)\cos z,$$

we have

$$|(j_n^{-1} \circ k_n)'(O)| = (|w_n - w_n'|/2)|\cos(k_n(O))||k_n'(O)|.$$

Since $K/4 < |f(z_n) - w_n| < K$ and $k_n(O) = H_n(f^*(z_n))$, we have that $|\cos(k_n(O))|$ is bounded away from O and also there exists a number $\lambda$ such that $|\mathrm{Im}(k_n(O))| \le \lambda$, where $\lambda$ is independent of n. However $d(n,\lambda) \to \infty$ as $n \to \infty$, and $|k_n'(O)| \ge d_2(O,k_n) \ge d(n,\lambda)$, and so $|k_n'(O)| \to \infty$ as $n \to \infty$. It follows that $(1 - |z_n|^2)|f'(z_n)| \to \infty$ as $n \to \infty$. But $f(z_n) \in N(E,K)$ for each n, and thus

$$\sup\{(1 - |z|^2)|f'(z)| : f(z) \in N(E,K)\} = \infty$$

in case (2) also. This completes the proof.

REFERENCES

1. P. Lappan, A Sufficient Condition for the Bloch Ordering, Bull. Inst. Math. Academia Sinica 10 (1982), 225-231.

2. ________ and L. Rubel, Some Orderings Induced by Spaces of Analytic Functions, Michigan Math. J. 27 (1980), 371-377.

3. W. Seidel and J.L. Walsh, On the Derivatives of Functions Analytic in the Unit Circle and Their Radii of Univalence and P-valence, Trans. Amer. Math. Soc. 52 (1942), 128-216.

Institute of Mathematics, Academia Sinica, Taipei, Taiwan
and
Department of Mathematics, Michigan State University
East Lansing, Michigan 48823

Department of Mathematics
Michigan State University
East Lansing, Michigan 48823

Contemporary Mathematics
Volume 25, 1983

ANALOGUES OF PICARD SETS FOR ENTIRE FUNCTIONS

AND THEIR DERIVATIVES

by

James K. Langley

## 1.  Introduction

A plane set E is a Picard set for entire functions if every transcendental entire function takes every finite complex value, with at most one exception, infinitely often in the complement of E.  It is well known that certain uncountable sets are Picard sets for entire functions--perhaps that most striking result in this direction is the following improvement by Anderson and Clunie [3], and Toppila [9], of an earlier result of Baker and Liverpool [4]:

### Theorem A

Suppose that $q > 1$ and $K > 0$ and that the complex sequence $(a_n)$ and the positive sequence $(\rho_n)$ satisfy, for all $n$,

$$|a_{n+1}| \geq q|a_n| \tag{1.1}$$

and

$$\log \frac{1}{\rho_n} \geq \frac{K}{\log q} (\log|a_n|)^2. \tag{1.2}$$

Then:

(a) if $K > \frac{1}{2}$, the union E of the discs $B(a_n,\rho_n) = \{z : |z - a_n| < \rho_n\}$ is a Picard set for entire functions:

(b) if $K \leq \frac{1}{2}$, there exists sequences $(a_n)$ and $(\rho_n)$ satisfying (1.1) and (1.2) such that the union $E_1$ of the discs $B(a_n,\rho_n)$ is not a Picard set for entire functions.

We shall be concerned with analogues of Theorem A for the value distribution of meromorphic functions and their derivatives.  Specifically if f is entire and transcendental, Hayman [6] proved that the equation $f^N(z)f'(z) = b$ must have infinitely many solutions, for any non-zero b, and integer N not less than two; Clunie [5] showed that the same conclusion holds for $N = 1$. The question then arises as to whether there are "exceptional" subsets of the plane outside which infinitely many of these points must lie, and in this direction, Anderson, Baker, and Clunie proved [2]:

J.K. LANGLEY

## Theorem B

Suppose that $a_n \to \infty$ such that, for all $n$,

$$\frac{a_{n+1}}{a_n} > q > 1.$$

Then if $f$ is a transcendental, entire function, and $N$ is an integer not less than two, the equation $f^N(z)f'(z) = b$ must have infinitely many solutions outside $E = \{a_n\}$, for any $b \neq 0$.

If $(a_n)$ satisfies

$$\liminf_{n \to \infty} \frac{|a_{n+1}|}{|a_n| \log |a_n|} > c > 0$$

the same conclusion holds for $N = 1$.

In the case where $N \geq 2$, we are able to extend Theorem B to obtain uncountable exceptional sets comparable to those of Theorem A, although our method fails for $N = 1$, due to difficulties in estimating the growth of $f$. We have

## Theorem 1

Suppose that the complex sequence $(a_n)$ and the positive sequence $(\rho_n)$ satisfy, for all n,

$$\frac{a_{n+1}}{a_n} > q > 1$$

and

$$\log \frac{1}{\rho_n} > \frac{q^{\frac{1}{4}} + 1}{q^{\frac{1}{4}} - 1} \; \frac{8}{\log q} \; (\log |a_n|)^2.$$

Then, if $f$ is a transcendental entire function, and $N$ is an integer not less than two, the equation $f^N(z)f'(z) = b$ must have infinitely many solutions outside the union of the discs $B(a_n, \rho_n)$, for any $b \neq 0$.

## Remark

This result was originally proved as part of the author's Ph.D. thesis, written under the supervision of I.N. Baker, to whom thanks are due.

We shall use the standard notation of Nevanlinna theory (see eg. [8]). If f is meromorphic in $|z| \leq r$, having $n(t,f)$ poles in $|z| \leq t$ (counting poles according to multiplicity), set

$$N(r,f) = \int_0^r (n(t,f) - n(0,f))\frac{dt}{t}$$

$$+ \, n(0,f)\log r$$

and

$$T(r,f) = \frac{1}{2\pi} \int_0^{2\pi} \log^+|f(re^{i\theta})|\,d\theta + N(r,f)$$

$$= m(r,f) + N(r,f)$$

where $\log^+ x = \max\{\log x, 0\}$. For $f$ analytic in $|z| \leq r$, we shall require

$$M(r,f) = \max\{|f(z)| \, : \, |z| = r\}$$

and the estimate

$$\log^+ M(s,f) \leq \frac{r+s}{r-s}\, T(r,f)$$

for $0 < s < r$. We shall use the term "nearly everywhere" (n.e.) as follows:
"n.e. as $r \to \infty$" will denote "as $r \to \infty$ outside a set of finite measure", while
if R is large, and S > R, "n.e. in $R \leq r < S$" will denote "for all r in
$R < r < S$ outside a set of small measure". We shall denote by $S(r,f)$ any
quantity satisfying $S(r,f) = o(T(r,f))$ n.e. as $r \to \infty$.

A key role will be played by the following result ([8], p. 57):

<u>Theorem C</u>

Suppose that $f(z)$ is meromorphic and non-constant in the plane,
and that

$$\psi(z) = \sum_{i=0}^{k} a_i(z) f^{(i)}(z)$$

is non-constant, where $k \geq 1$ and, for each i,

$$T(r,a_i) = S(r,f).$$

Then

$$T(r,f) < \bar{N}(r,f) + N(r, \frac{1}{f})$$

$$+ \, \bar{N}(r, \frac{1}{\psi - 1}) - N_0(r, \frac{1}{\psi'}) + S(r,f)$$

where $N_0(r, \frac{1}{\psi'})$ counts only zeros of $\psi'$ which are not roots of $\psi(z) = 1$, and
$\bar{N}(r,f)$, $\bar{N}(r, \frac{1}{\psi - 1})$ count only points at which $f = \infty$, $\psi = 1$ respectively.
without regard to multiplicity.

We shall require the following result of Hayman [7]:

Theorem D
>
> Suppose that $f(z)$ is entire and satisfies
>
> $$T(r,f) = 0(\log r)^2.$$

Then

$$\log\left|f(re^{i\theta})\right| \sim \log M(r,f)$$

as $z = re^{i\theta}$ tends to infinity outside an $\varepsilon$-set surrounding the zeros of $f$.

The term "$\varepsilon$-set" denotes a countable set of discs not meeting the origin, which subtend angles at the origin whose sum is finite.

## 2.  Preliminary Lemmas

Lemma 1
>
> Suppose that $P(z)$ is a polynomial of degree $k$ with all its zeros $b_1,\ldots,b_k$ lying in $|z| < R_0$.  Then, for $|z| = R \geq R_0$,
>
> $$\left|\frac{P'(z)}{P(z)}\right| > \frac{k}{2R}.$$

Proof

We have

$$\frac{P'(z)}{P(z)} = \sum_{i=1}^{k} \frac{1}{z - b_i} = \frac{1}{z} \sum_{i=1}^{k} \left(1 - \frac{b_i}{z}\right)^{-1}$$

Now, if $|\omega| < 1$, $\mathrm{Re}\, \frac{1}{1 - \omega} > \frac{1}{2}$, since $t = (1 - \omega)^{-1}$ maps the unit disc $|\omega| < 1$ into the half-plane $\mathrm{Re}(t) > \frac{1}{2}$.  Thus, if $|z| = R \geq R_0$, then $\left|\frac{b_i}{z}\right| < 1$ and so

$$\left|\frac{P'(z)}{P(z)}\right| = \frac{1}{R}\left|\sum_{i=1}^{k} \left(1 - \frac{b_i}{z}\right)^{-1}\right|$$

$$\geq \frac{1}{R}\left|\mathrm{Re} \sum_{i=1}^{k} \left(1 - \frac{b_i}{z}\right)^{-1}\right|$$

$$> \frac{k}{2R}.$$

Lemma 2
>
> Suppose that $h(z)$ is regular and non-zero in $|z| \leq R$, and that $\left|\log|h(z)|\right| \leq M$ on $|z| = R$.  Then for $|z| = r < R$,
>
> $$\left|\frac{h'(z)}{h(z)}\right| \leq \frac{2MR}{(R - r)^2}$$

Proof

We have (see eg. [8], p. 22)

$$\frac{h'(z)}{h(z)} = \frac{1}{\pi} \int_0^\pi \log|h(Re^{i\phi})| \frac{Re^{i\phi}}{(Re^{i\phi} - z)^2} \, d\phi$$

and simple estimates yield Lemma 2.

3. **Proof of Theorem 1**

Suppose that there exists a transcendental entire function $f(z)$, having only finitely many solutions of $f^N(z)f'(z) = b$ outside the union of the discs $D_n = B(a_n, \rho_n)$ such that,

$$\frac{a_{n+1}}{a_n} > q > 1 \tag{3.1}$$

and

$$\log \frac{1}{\rho_n} > \frac{q^{\frac{1}{4}} + 1}{q^{\frac{1}{4}} - 1} \frac{8}{\log q} (\log|a_n|)^2 \tag{3.2}$$

while $N$ is an integer not less than two, and $b \neq 0$. It is clear that we may assume that $b = 1$, since otherwise we need only consider $g(z) = \alpha f(z)$ where $\alpha^{N+1} = b^{-1}$. We shall show by a series of steps that $f$ is of very small growth, and finally obtain a contradiction. We set

$$F(z) = \frac{1}{N+1} f^{N+1}(z), \tag{3.3}$$

and note that all large 1-points of $F'$ lie in the discs $D_n$. Applying Theorem C (with $\psi = F'$), we have

$$T(r,f) < N(r, \frac{1}{F}) + \bar{N}(r, \frac{1}{F' - 1})$$

$$- N_0(r, \frac{1}{F''}) + S(r,F)$$

where $N_0(r, \frac{1}{F''})$ counts only zeros of $F''(z)$ which are not 1-points of $F'(z)$, and $S(r,F) = o(T(r,F))$ n.e. as $r \to \infty$. Hence

$$T(r,F) < N(r, \frac{1}{F}) + N(r, \frac{1}{F' - 1})$$

$$- N(r, \frac{1}{F''}) + S(r,F) \tag{3.4}$$

Now, zeros of $F$ are zeros of $f$, and thus have multiplicity at least 3, from (3.3). Thus, if $\hat{N}(r, \frac{1}{F''})$ counts the zeros of $F''$ which are not zeros of $F$, and which lie in the discs $D_n$, we have

$$N(r, \frac{1}{F}) - N(r, \frac{1}{F''}) \leq 2 \bar{N}(r, \frac{1}{F}) - \hat{N}(r, \frac{1}{F''}) , \tag{3.5}$$

where $\bar{N}(r, \frac{1}{F})$ counts the points at which $F = 0$, without regard to multiplicity.

80                        J.K. LANGLEY

But then

$$\bar{N}(r, \tfrac{1}{F}) \leq \frac{1}{N+1} N(r, \tfrac{1}{F}) \leq \tfrac{1}{3} M(r,F) + 0(1)$$

and so (3.4) and (3.5) yield

$$T(r,F) < 4 N\,(, \frac{1}{F'-1}) - \hat{N}(r, \tfrac{1}{F''}) \tag{3.6}$$

n.e. as $r \to \infty$.

We define sequences $p_n$, $t_n$, and $v_n$ as follows. In both cases counting multiplicities, let $p_n$ be the number of 1-points of $F'$ in the disc $D_n$, and $t_n$ the number of zeros of $F''$ in $D_n$ which are not zeros of $F$, and set

$$v_n = p_n - t_n. \tag{3.7}$$

Then for large M, we have n.e. in $q^{\frac{1}{2}}|a_M| \leq r \leq |a_{M+1}| - 1$,

$$\tfrac{1}{4} T(r,F) < 0(\log r) + \sum_{n=1}^{M} p_n\left(\log \frac{r}{|a_n|} + o(1)\right)$$

$$- \sum_{n=1}^{M} t_n\left(\log \frac{r}{|a_n|} + o(1)\right). \tag{3.8}$$

Now,

$$\sum_{n=1}^{M} (p_n + t_n) \leq n(|a_M|+1, \frac{1}{F'-1}) + n(|a_M|+1, \tfrac{1}{F''})$$

$$= 0(T(r,F) + S(r,F)) \tag{3.9}$$

for $r \geq q^{\frac{1}{2}}|a_M|$. Since F is transcendental, by the assumption on f, we have, from (3.8) and (3.9),

$$T(r,F) < \frac{9}{2} \sum_{n=1}^{M} v_n \log \frac{r}{|a_n|} \tag{3.10}$$

n.e. in $q^{\frac{1}{2}}|a_M| \leq r \leq |a_{M+1}| - 1$, if M is large enough. Now, whether or not the sequence $(v_n)$ is bounded above, there must exist $m_0$ and infinitely many M such that

$$v_M = \max\{v_m : m_0 < m \leq M\} \geq 1 \tag{3.11}$$

since otherwise F would not be transcendental.

Suppose now that M is large, and satisfies (3.11). Then n.e. in $a^{\frac{1}{2}}|a_M| \leq r \leq |a_{M+1}| - 1$ we have, from (3.10),

$$T(r,F) < 0(\log r) + \frac{9}{2} v_M \sum_{m=m_0}^{M} \log \frac{r}{|a_m|} . \tag{3.12}$$

But

$$\sum_{m=1}^{M} \log \frac{r}{|a_m|} \quad M \log \frac{r}{|a_1|}$$

$$= M(1 + o(1))\log r$$

$$\leq (1 + o(1))(\log q)^{-1}(\log r)^2 \tag{3.13}$$

for M and r as above, since

$$\log r \geq \log q^{\frac{1}{2}}|a_M| \geq (M - \frac{1}{2})\log q + \log|a_1| .$$

Thus, from (3.12),

$$T(r,F) < \frac{5}{\log q} v_M(\log r)^2 \tag{3.14}$$

n.e. in $q^{\frac{1}{2}}|a_M| \leq r \leq |a_{M+1}| - 1$, for any large M satisfying (3.11).

We go on to show that if M is large, and satisfies (3.11), then F has no zeros in $D_M$, and $v_M \leq 1$. From (3.14).

$$T(r,F' - 1) = T(r,F') + O(1)$$

$$< (1 + o(1))T(r,F)$$

$$< \frac{6}{\log q} v_M(\log r)^2 \tag{3.15}$$

n.e. in $q^{\frac{1}{2}}|a_M| \leq r \leq |a_{M+1}| - 1$. In particular, if M is large and satisfies (3.11),

$$T(q^{\frac{1}{2}}|a_M|,F' - 1) < \frac{7}{\log q} v_M(\log|a_M|)^2 \tag{3.16}$$

since (3.14) holds for some r with $q^{1/2}|a_M| \leq r \leq q^{3/4}|a_M|$. Now suppose that the 1-points of F' in $D_M$ are $c_1,\ldots,c_{p_M}$. We set

$$P(z) = \prod_{k=1}^{p_M} (z - c_k) \tag{3.17}$$

and

$$F'(z) - 1 = h(z)P(z) \tag{3.18}$$

On the circle $|z - a_M| = 4$,

$$\log|P(z)| > p_M \log 3$$

82

J.K. LANGLEY

(noting that $p_M \geq v_M \geq 1$) while

$$\log|F'(z) - 1| \leq \log M(|a_M| + 4, F' - 1)$$

$$\leq \frac{q^{\frac{1}{4}} + 1}{q^{\frac{1}{4}} - 1} \; T(q^{\frac{1}{2}}|a_M|, F' - 1). \tag{3.19}$$

Thus, on $|z - a_M| = 4$, from (3.19),

$$\log|h(z)| < \frac{q^{\frac{1}{4}} + 1}{q^{\frac{1}{4}} - 1} \; T(q^{\frac{1}{2}}|a_M|, F' - 1)$$

$$- p_M \log 3$$

and the same estimate holds in $|z - a_M| < 4$, by the maximum principle. But

for $z$ in the disc $D_M$, we have $|z - c_k| < 2\rho_M$, and thus using (3.16),

$$\log|F'(z) - 1| = \log|P(z)| + \log|h(z)|$$

$$< p_M \log 2 - p_M \log 3$$

$$+ \frac{q^{\frac{1}{4}} + 1}{q^{\frac{1}{4}} - 1} \; \frac{7}{\log q} v_M (\log|a_M|)^2$$

$$< p_M \left( \frac{q^{\frac{1}{4}} + 1}{q^{\frac{1}{4}} - 1} \; \frac{8}{\log q} (\log|a_M|)^2 \right)$$

$$- p_M \log \frac{1}{\rho_M}$$

$$< 0,$$

by (3.2), provided M is large enough. In particular, $F'(z) \neq 0$ in $D_M$. Since

zeros of F are zeros of F', we conclude that F cannot have any zeros in $D_M$, if

M is large and satisfies (3.11).

To show that $v_M \leq 1$, we note that, if M is large enough,

$$T(q^{\frac{1}{2}}|a_M|, h) \leq T(q^{\frac{1}{2}}|a_M|, F' - 1) + T(q^{\frac{1}{2}}|a_M|, \frac{1}{P})$$

$$\leq T(q^{\frac{1}{2}}|a_M|, F' - 1) + T(q^{\frac{1}{2}}|a_M|, P)$$

$$\leq \frac{7}{\log q} v_M (\log|a_M|)^2 + p_M \log 2q^{\frac{1}{2}}|a_M|$$

$$\leq \frac{8}{\log q} p_M (\log a_M)^2 \tag{3.20}$$

using (3.16), and noting that $|P(0)| > 1$.

We estimate $\left|\log|h(z)|\right|$ for $z = re^{i\theta}$ lying in $|z - a_M| \leq 4$.

The Poisson-Jensen formula, applied to h in $|\omega| < R = q^{\frac{1}{2}}|a_M|$ yields

$$\log|h(z)| = \frac{1}{2\pi} \int_0^{2\pi} \log|h(Re^{i\theta})| \; \frac{(R^2 - r^2)d\phi}{R^2 + r^2 - 2Rr\cos(\theta - \phi)}$$

$$+ \sum_{\substack{\xi \,=\, \text{zeros of} \\ h \text{ in } |\omega| < R}} \log \left| \frac{R^2 - \bar{\xi}z}{R(\xi - z)} \right| \tag{3.21}$$

Now, any zero, $\xi$ say, of h in $|\omega| < R$ is a zero of $F' - 1$, and thus $|\xi| < |a_{M-1}| + 1$, and so $|\xi - z| \geq \frac{1}{2}(1 - \frac{1}{q})|a_M|$, provided M is large enough. Thus (3.12) yields

$$\left|\log|h(z)|\right| \leq \frac{R + r}{R - r} \left( m(r,h) + m(R, \frac{1}{h}) \right)$$

$$+ n\left(|a_{M-1}| + 1, \frac{1}{F' - 1}\right) \log \frac{4R}{(1 - \frac{1}{q})|a_M|}$$

$$\leq \hat{c}_1 (m(R,h) + m(R, \frac{1}{h}))$$

$$+ \hat{c}_2 \; T(R, F' - 1) \tag{3.22}$$

where we use $\hat{c}_1, \hat{c}_2, \hat{c}_3, \ldots$ to denote constants depending only on q. But

$$m(R, \frac{1}{h}) \leq m(R, \frac{1}{F' - 1}) + m(R, P)$$

$$\leq T(R, F' - 1) + p_M \log 2R$$

and thus (recalling that $R = q^{\frac{1}{2}}|a_M|$) we have, from (3.16), (3.20) and (3.22)

$$\left| \log|h(z)| \right| \leq \hat{c}_3 \; p_M (\log|a_M|)^2$$

for $|z - a_M| \leq 4$. Thus by Lemma 2,

$$\left| \frac{h'(z)}{h(z)} \right| \leq \hat{c}_4 \; p_M (\log|a_M|)^2$$

for $|z - a_M| \leq 2$. But P(z) has all its zeros in the disc $D_M$, and so, by Lemma 1,

$$\left| \frac{P'(z)}{P(z)} \right| > \frac{2}{\rho_M} \; p_M$$

on $|z - a_M| = \rho_M$, and hence, using (3.2)

$$|P'(z)h(z)| > |P(z)h'(z)|$$

on $|z - a_M| = \rho_M$, if M is large enough.  But, from (3.18),

$$F'(z) = P'(z)h(z) + P(z)h'(z)$$

and so, by Rouché's Theorem, F" has the same number of zeros in $|z - a_M| < \rho_M$
as P'(z)h(z).  But h(z) does not vanish in $D_M$, while P'(z) has $p_M - 1$ zero,
which all lie in the convex hull of the set of zeros of P(z) ([1], p. 29) and
hence all lie in $D_M$.  So F" has $p_M - 1$ zeros in $D_M$, and since, as we saw
earlier, F does not vanish in $D_M$, we have

$$t_M = p_M - 1$$

for all large enough M satisfying (3.11); hence, for such M,

$$v_M = p_M - t_M = 1$$

and thus $v_n \leq 1$ for all large n.

Returning to the estimate (3.10), we now have

$$T(r,F) < \frac{9}{2} \sum_{n=1}^{M} \log \frac{r}{|a_n|}$$

and hence, using (3.13),

$$T(r,F) < \frac{5}{\log q} (\log r)^2 \tag{3.23}$$

n.e. in $q^{\frac{1}{2}}|a_M| \leq r \leq |a_{M+1}| - 1$, if M is large enough, and hence

$$T(r,F) = O(\log r)^2$$

for all large r; in particular, F has order zero, and so (3.23) holds for
$r = q^{\frac{1}{2}}|a_M|$, if M is large.  Moreover,

$$T(r,F' - 1) \leq (1 + o(1))T(r,F)$$
$$= O(\log r)^2 \tag{3.24}$$

and, for large M,

$$T(q^{\frac{1}{2}}|a_M|,F' - 1) < (1 + o(1)) \frac{5}{\log q} (\log q^{\frac{1}{2}}|a_M|)^2$$
$$< \frac{6}{\log q} (\log|a_M|)^2. \tag{3.25}$$

We are now in a position to obtain a contradiction.  We take a positive
$\varepsilon$ so small that the discs $B(a_n, 4\varepsilon|a_n|)$ are disjoint.  Since all large zeros
of F' - 1 lie in the discs $D_n = B(a_n, \rho_n)$ we see from Theorem D that

$|F'(z) - 1| > 10$, say, for large $z$ outside the union of the discs $B(a_n, \varepsilon|a_n|)$.

Now consider a large zero, $z_0$, of $f(z)$ (note that since $F$ has order zero but is assumed transcendental, $F$ and hence $f$ must have infinitely many zeros). Since $F'(z) = f^N(z) f'(z)$ we have $F'(z_0) = 0$, and hence $z_0$ must lie in some disc $B(a_M, \varepsilon|a_M|)$; but then, by Rouche's Theorem, $F'$ must have a 1-point, $\omega_0$ say, in $B(a_M, \varepsilon|a_M|)$, which must lie in the smaller disc $D_M$.

We show now that

$$\log|F'(z) - 1| < 0 \text{ in } D_M. \tag{3.26}$$

On the circle $|z - a_M| = 4$, we have $|z - \ddot{\omega}_0| > 3$, while

$$\log|F'(z) - 1| \leq \log M(|a_M| + 4, F' - 1)$$

$$\frac{q^{\frac{1}{4}} + 1}{q^{\frac{1}{4}} - 1} \frac{6}{\log q} (\log|a_M|)^2$$

using (3.25). But then, by the maximum principle,

$$\log\left|\frac{F'(z) - 1}{z - \omega_0}\right| < \frac{q^{\frac{1}{4}} + 1}{q^{\frac{1}{4}} - 1} \frac{6}{\log q} (\log|a_M|)^2$$

$$- \log 3$$

for $|z - a_M| \leq 4$, and so, for $|z - a_M| < \rho_M$,

$$\log|F'(z) - 1| < \frac{q^{\frac{1}{4}} + 1}{q^{\frac{1}{4}} - 1} \frac{6}{\log q} (\log|a_M|)^2$$

$$- \log 3 + \log(2\rho_M)$$

$$< 0$$

using (3.2). Thus (3.26) holds, and $\mathrm{Re}(F'(z) > 0$ in $D_M$.

In particular, $z_0$ cannot lie in $D_M$, since $F'(z_0) = 0$; however, $z_0$ lies in a component of $\{z : |F'(z)| < 1\}$ whose boundary is a level curve $\Gamma$, contained entirely in $B(a_M, \varepsilon|a_M|)$, on which $|F'(z)| = 1$. Since $\Gamma$ must close in $B(a_M, \varepsilon|a_M|)$, $\Gamma$ must pass through a 1-point of $F'(z)$ and hence must meet $D_M$. Consider now a ray from $z_0$ which does not meet the disc $D_M$, and suppose that the first point of intersection of this ray with $\Gamma$ is at $z_1$, say. Starting from $z_1$ we may pass along $\Gamma$ in either direction until the first point of intersection with the closed disc $\bar{D}_M$, and suppose that these points are

$z_2$, $z_3$, say.  Then we may form a closed curve J, consisting of the line-segment from $z_2$ to $z_3$ and the arc of $\Gamma$ which contains $z_1$, $z_2$, and $z_3$ and in whose interior lies $z_0$.  We consider arg $F'(z)$ on this composite curve J.

On the open line-segment from $z_2$ to $z_3$ (which is contained in $D_M$) we have $Re(F'(z)) > 0$ and so arg $F'(z)$ cannot change by more than $\pi$ as z moves from $z_2$ to $z_3$ along this line-segment.  On the arc of $\Gamma$ which contains $z_1$, $z_2$, and $z_3$ (and which forms the other component of J) we have $|F'(z)| = 1$ but this arc passes through no 1-points of $F'(z)$.  Hence arg $F'(z)$ cannot change by more than $2\pi$ as z moves from $z_3$ to $z_2$ along this arc.  We conclude that arg $F'(z)$ cannot increase by more than $3\pi$ as z describes the closed curve J once (in either direction) and hence by the argument principle $F'(z)$ can have at most one zero in the interior of the curve J.  But $z_0$ lies in this interior, and is a zero of $F'(z)$ of multiplicity at least N, i.e. at least 2.  Thus we have obtained a contradiction and conclude that f(z) cannot be transcendental.

## REFERENCES

[1]  Ahlfors, L.V., "Complex Analysis", McGraw-Hill, 1966.

[2]  Anderson, J.M., Baker, I.N. and Clunie, J., "The Distribution of Values of Certain Entire and Meromorphic Functions", Math. Zeit., Band 178, Heft 4, (1981).

[3]  Anderson, J.M. and Clunie, J., "Picard Sets of Entire and Meromorphic Functions", Ann. Acad. Sci. Fenn. Ser. Al5 (1980), 27-43.

[4]  Baker, I.N. and Liverpool, L.S.O., "Picard Sets for Entire Functions", Math. Zeitschrift 126 (1972), 230-238.

[5]  Clunie, J., "On a Result of Hayman", Journal of the London Math. Soc. 47 (1967), 389-392.

[6]  Hayman, W.K., "Picard Values of Meromorphic Functions and Their Derivatives", Ann. of Math. (2), 70 (1959), 9-42.

[7]  Hayman, W.K., "Slowly Growing Integral and Subharmonic Functions", Comment. Math. Helv. 34 (1960), 75-84.

[8]  Hayman, W.K., "Meromorphic Functions", Oxford, 1964.

[9]  Toppila, S., "On the Value Distribution of Meromorphic Functions with a Deficient Value", Ann. Acad. Sci. Fenn., Ser. Al5 (1980), 179-184.

J.K. LANGLEY
Department of Mathematics
Imperial College
London, England
and

DEPARTMENT OF MATHEMATICS
UNIVERSITY OF ILLINOIS
URBANA, ILLINOIS

Contemporary Mathematics
Volume **25**, 1983

## EXCEPTIONAL SETS FOR ENTIRE FUNCTIONS

### BY

Lennox S.O. Liverpool   and Kenneth B. Yuguda

## INTRODUCTION

Following LEHTO [6] we call a plane set E a Picard set for entire functions if every entire transcendental function $f(z)$ has the following property.  For every finite value a, with at most one exception, $f(z)$ takes a infinitely often in the complement CE of E.  This is the only sense in which we shall use the expression "Picard set" in what follows.

From this definition, if E is not a Picard set, there must be an entire transcendental function $f(z)$, such that for two distinct values a and b, $f(z)$ takes a and b at most finitely often in CE.  Clearly we can assume without loss of generality that a and b are 0 and 1 respectively.

Some results on Picard sets in this sense may be found in Toppila [11] and Baker and Liverpool [2] and [3] and in Liverpool and Umar [8].  In Lehto [6] the following result was proved.

A set $E = \{a_n\}$ $n = 1,2,3, \ldots$ is a Picard set for entire functions if

$$\left| \frac{a_n}{a_{n+1}} \right| = 0(1/n^2)$$

By restricting the points of E to lie on a ray the above condition can be weakened to

$$\left| \frac{a_{n+1}}{a_n} \right| \geq q > 1$$

In Liverpool and Umar [8], we proved in a direct manner that the restriction "points of E lie on a ray" is not necessary.  Indeed the condition

$$\left| \frac{a_{n+1}}{a_n} \right| \geq q > 1$$

is sufficient to ensure our point set $E = \{a_n\}$ is a Picard set.

We observed also that the entire function $f(z) = \tfrac{1}{2}(1+\cos\sqrt{z})$

takes 0 and 1 at the points $a_n = n^2\pi^2$, $n = 0, 1, 2, \ldots$.

This shows that the condition

$$\left| \frac{a_{n+1}}{a_n} \right| > 1 + 2/n$$

is not strong enough to give us Picard sets.  However, for this function, and its zero and one points,

$$\left|\frac{a_{n+1}}{a_n}\right| \to 1 \text{ as } n \to \infty$$

We showed that we can indeed obtain Picard set $E = \{a_n\}$ for which

$$\left|\frac{a_{n+1}}{a_n}\right| \to 1 \text{ as } n \to \infty$$

More precisely, we proved the following result:

THEOREM A.

If $E = \{a_n\}$ $n = 1,2,3, \ldots,$ is a countable point set and

$$\left|\frac{a_{n+1}}{a_n}\right| \geq q > 1$$

Then $E = \{a_n\}$ is a Picard set for entire transcendental functions.

Moreover, we can obtain Picard sets $E = \{a_n\}$ for which

$$\left|a_n\right| < \left|a_{n+1}\right| \to \infty$$

and

$$\left|\frac{a_{n+1}}{a_n}\right| \to 1 \text{ as } n \to \infty$$

In particular any set with

$$\left|a_n\right| = \exp\left(\frac{\delta n}{\log n}\right), \ \delta > \pi^2$$

in such a set.

After establishing this theorem, we briefly discussed our result in the light of other related results in the literature.  In particular we showed that with our techniques we could easily derive:

THEOREM B - A result of Matsumoto [9].

If $f(z)$ is an entire transcendental function, then $f(z)$ takes every finite value infinitely often in CE, with at most one exception if the points $\{a_n\}$ of E satisfy the condition

$$\log \left|\frac{a_{n+1}}{a_n}\right| \geq m(n)$$

where $m(n)$, $n = 1,2,3, \ldots$ are positive numbers such that

$$\varlimsup_{t\to\infty} \frac{k^{\frac{1}{m(t)}}}{\sum\limits_{n=1}^{t} m(n)} < \infty$$

where k is a constant

We shall in this paper improve the result of theorem A in two directions and then discuss the application of an exceptional set result to a problem in factorization of entire functions.

## STATEMENT AND PROOF OF RESULTS

THEOREM 1.

If there are constants q,k both greater than unity such that

    (i)   for all sufficiently large r, the annulus $r/q < |z| < q r$ contains at most two members of a set E

    (ii)  The only point of E contained in $|z-a_n| < \dfrac{1}{|a_n|^k}$

         is $a_n$ then E is a Picard set for entire functions.

PROOF:

Suppose E with conditions above is not a Picard set, then there is an entire transcendental function f(z) whose zeros and ones are in E.  For such f(z)

$$T(r,f) = 0(\log r)^2 = \log M(r,f)$$

In this case (c.f. eg Boas [4]) for any $\varepsilon > 0$, and for any $q > \lambda' > \lambda > 1, \exists$ for all large enough r, numbers r', r" with

$$\frac{r}{\lambda'} < r' < \frac{r}{\lambda} < r < \lambda r < r'' < \lambda' r'$$

such that

$$m(r',f) > M(r',f)^{1-\varepsilon} > 1$$

$$m(r'',f) > M(r'',f)^{1-\varepsilon} > 1$$

We choose $r = |b_n|$, $f(b_n) = 0$ and $|b_n|$ large; and define r', r" as above. Rouche's theorem gives, f(z) and f(z)-1 have the same number of zeros in

$$A : r' < |z| < r'' ,$$

Since there are at most two members of E in the annulus, it follows we must have exactly two, $a_n$ with $f(a_n) = 1$ and $b_n$ with $f(b_n) = 0$, of the same multiplicity $m_n$.  In fact each $b_n$ is embedded in such an annulus.  If we now suppose all large $b_n$ have multiplicity $m_n \geq k$, Nevanlinna's theory on deficient values gives $k \leq 2$.

Consider now $\dfrac{f(z)}{(z-b_n)^k}$ which is regular and non-zero in the annulus between $|z| = r'$ and $|z| = r''$ and on these circles.  Since we may suppose $b_n$ and hence r', r" so large that for any constant B > 2k in (ii) of the theorem

$$M(r',f) > \{4\lambda'^{4+2B} r'^{(B+2)}\}^{1/1-\varepsilon}$$

We can show that throughout the annulus

$$\left|\frac{f(z)}{(z-b_n)^k}\right| \geq (r'')^B$$

In particular at $z = a_n$

$$\frac{1}{|a_n - b_n|^k} \geq (r'')^B > |a_n|^B$$

which gives

$$|a_n - b_n| < \frac{1}{|a_n|} \quad \text{contradicting hypothesis (ii),}$$

since $B > 2k$

THEOREM 2.

If $a_n = \lambda_n e^{n^{1/p}}$, $n = 1,2,3, \ldots$ where $\lambda_n > 0$, $\lambda_n$ is non-decreasing and if $p \leq 2$, then $E = \{a_n\}$ is a Picard set for entire transcendental functions.

REMARKS.

Without loss of generality we can assume $\lambda_1 \geq 1$. It suffices to discuss the case $p > 1$ since $p \leq 1$ is already covered by Theorem A.

We note in particular that if we choose $\lambda_n = 1$, $p = 2$, $E = \{a_n\} = \{e^{\sqrt{n}}\}$, a function which has all its zeros and ones at these points would have its Nevanlinna characteristic $T(r)$ satisfying

$$T(r) = \frac{1}{3}(\log r)^3$$

compared with

$$T(r) = 0(\log r)^2 \quad \text{in Theorem A}.$$

In general our theorem gives

$$T(r) = 0(\log r)^{p+1}$$

and for $2 \geq p > 1$ we have an improvement of the earlier result.

PROOF.

If $E = \{a_n\}$ as specified in the statement of the theorem above is not a Picard set for entire functions, then there exists an entire function $f(z)$ with all its zeros and ones at points in E.

Such $f(z)$ necessarily have

$$T(r,f) = 0(\log r)^{p+1}$$

and must be of order zero.

We can thus write

$$f(z) = k\prod_{n=1}^{\infty} (1 - z/a_n)^{\mu'_n}$$

where $\{\alpha_n\} \subset \{a_n\}$, is the set of zeros of $f(z)$. Since there are points $d_n$ which belong to the sequence $\{a_n\}$ such that $f(d_n) = 1$, we see that $k$ is real. We can now show that $\mu_n' = 1$ or $2$. We take a component $C$ of $|f(z)| < 1$ containing $\alpha_n$, $f(z)$ is real on the real axis, so $C$ is symmetric about it, simply connected and bounded by a simple closed curve $\Gamma$ on which $|f(z)| = 1$. $\Gamma$ meets $R$, the real line only in two points. If $\mu_{n'} > 2$, $f(z)$ must take $1$, $\mu_{n'}(> 2)$ times on $\Gamma$, which means the value $1$ is taken at some non-real point as well as the two points where $\Gamma$ meets $R$, contradicting the assumption that all "ones" are in $E = \{a_n\}$.

We employ $a_n$ in our factorization by writing

$$f(z) = k \prod_{n=1}^{\infty} (1 - {}^z\!/a_n)^{\mu_n} \text{ with } \mu_n = 0, 1, \text{ or } 2$$

with the understanding that if $a_n$ is not one of the zeros, $\mu_n = 0$.

We define $P_n = \prod_{k=n+1}^{\infty} (1 - \dfrac{a_n}{a_k})$ and show $1 > P_n > \dfrac{1}{a_n^c}$, $c > 16$.

Using $P_n$ as defined above,

$$0 > \log P_n > \sum_{k=n+1}^{\infty} \log (1 - \frac{a_n}{a_k})$$

$$= \sum_{k=n+1}^{\infty} \log[1 - \frac{\lambda_n}{\lambda_k} \exp (n^{1/p} - k^{1/p})]$$

$$\geq \sum_{k=n+1}^{\infty} \log[1 - \exp (n^{1/p} - k^{1/p})] = T_n$$

We introduce

$$\phi(x) = \log [1 - \exp (n^{1/p} - x^{1/p})]$$

which increases to zero.

Then $\displaystyle\sum_{k=n+1}^{\infty} \phi(k) = T_n > \int_n^{\infty} \phi(x)\,dx$

$$= \int_n^{\infty} \log[1 - \exp (n^{1/p} - x^{1/p})]\,dx = J \text{ say}$$

We make a change of variable to $t$ by employing the substitution

$$n^{1/p} - x^{1/p} = -t$$

Thus $J = p \displaystyle\int_0^{\infty} \{\log(1 - e^{-t})\}\, (n^{1/p} + t)^{p-1}\, dt$

J converges and we estimate it as follows

$$-J = p \int_0^{n^{1/p}} \{-\log(1-e^{-t})\}(n^{1/p} + t)^{p-1} dt$$

$$+ p \int_{n^{1/p}}^{\infty} \{-\log(1 - e^{-t})\}(n^{1/p} + t)^{p-1} dt$$

$$< p \int_0^{n^{1/p}} \{-\log(1 - e^{-t})\}(2n^{1/p})^{p-1} dt$$

$$+ p \int_{n^{1/p}}^{\infty} \{-\log(1 - e^{-t})\}(2t)^{p-1} dt$$

$$< 2^{p-1}, p.n^{\frac{p-1}{p}} \int_0^{\infty} \{-\log(1 - e^{-t})\} dt$$

$$+ 2.p \int_0^{\infty} \{-\log(1 - e^{-t})\} t^{p-1} dt$$

$$= A\, n^{\frac{p-1}{p}} + B < C\, n^{\frac{p-1}{p}} \quad \text{for a suitable constant C.}$$

We choose a C for which this holds and assume from now on that $C > 16$.

Thus $0 > \log P_n > J > -\, Cn^{\frac{p-1}{p}}$

which means

$$1 > P_n > \exp(-Cn^{\frac{p-1}{p}})$$

But $1 < p \leq 2$ implies $\dfrac{p-1}{p} < \dfrac{1}{p}$ and we get

$$1 > P_n > \exp(-C\, n^{1/p}) \geq \frac{1}{\lambda_n^C} \exp(-C\, n^{1/p})$$

since $\lambda_n \geq \lambda_1 \geq 1$, and thus

$$1 > P_n > \frac{1}{a_n^C} \quad \text{as required.}$$

We now return to the consideration of our $f(z)$ which is of zero order. For any $\varepsilon > o$, and all large enough R, there exists in $(R, R^{1+\varepsilon})$ values of r where the minimum modulus $m(f,r)$ satisfies $m(f,r) > r^{3c}$ {see e.g. Baker [1]}. Hence taking an arbitrarily large $a_m$ which is a zero of $f(z)$, there is such an r with $a_m > r > a_m^{3/4}$ and if $a_n$ is the first zero of $f(z)$ larger than r, we have

$$\left.\begin{array}{l} m(f,r) > r^{3c} \\[4pt] a_n^{3/4} \leq a_m^{3/4} < r < a_n \\[4pt] f(a_n) = 0 \end{array}\right\}$$

We assume $a_m$ and thus $r$ chosen so large that $f(z)$ has more than $3c$ zeros in the

interval $(0,r)$.

We now put

$$Q(z) = k \prod_{|a_m| < r} (1 - \frac{z}{a_m})^{\mu_m}$$

$$R(z) = \prod_{m=n+2}^{\infty} (1 - \frac{z}{a_m})^{\mu_m}$$

So that $f(z) = Q(z) (1 - \frac{z}{a_n})^{\mu_n} (1 - \frac{z}{a_{n+1}})^{\mu_{n+1}} R(z)$

with $\mu_n = 1$ or $2$, $\mu_{n+1} = 0, 1,$ or $2$

Now $m(f,r) = f(r) = Q(r) (1 - \frac{r}{a_n})^{\mu_n} (1 - \frac{r}{a_{n+1}})^{\mu_{n+1}} R(r) > r^{3c}$

when $Q(r) = m(Q,r) > r^{3c}$.

Since $Q(z)$ has more than $3c$ zeros, we see that

$$\left| \frac{Q(z)}{z^{3c}} \right| \to 0 \text{ as } |z| \to \infty \text{ and so by the}$$

maximum modulus theorem,

$$|Q(z)| > |z|^{3c} \text{ for all } |z| \geq r$$

Now if $g(z) = Q(z)R(z)$, we have on $|z| = r$

$$|g(z)| \geq g(r) > f(r) > r^{3c} > a_n^{c}$$

while on $|z| = a_{n+1}$ we have

$$|g(z)| > |Q(a_{n+1}) R(a_{n+1})|$$

$$\geq |a_{n+1}|^{3c} \prod_{m=n+1}^{\infty} (1 - \frac{a_{n+1}}{a_m})^{\mu_m}$$

$$\geq |a_{n+1}|^{3c} \cdot p_{n+2}^{2}$$

$$\geq |a_{n+1}|^{3c} \cdot \frac{1}{|a_{n+1}|^{2c}} = |a_{n+1}|^{c}$$

Thus on the boundary of the annulus $A: r \leq |z| \leq |a_{n+1}|$ and hence throughout

$A$, the regular non-vanishing function $g(z)$ satisfies

$$|g(z)| > |a_n|^c$$

and

$$|f(z)| > |a_n|^c \, (1 - \frac{z}{a_n})^{\mu_n} (1 - \frac{z}{a_{n+1}})^{\mu_{n+1}}$$

Now by Laguerre's theorem (Holland [5]) an entire function $F(z)$ of order less than 2, real on the real axis and having all its zeros real is such that the zeros of $F'(z)$ are all real and separated from each other by the zeros of $F(z)$. Applied to $f(z)$ and $f(z)-1$ in our case, we see that the zeros of $f'(z)$ are real and separated by the zeros of $f(z)$ and also by the "ones" of $f(z)$. Thus $f(z)$ cannot be negative immediately on both sides of a zeros, $a_n$ of $f(z)$, for if not we would have two zeros of $f'(z)$ not separated by a "one" of $f$. Now from our proceeding estimate for $f(z)$ in A we have

$$\left|f(\frac{a_n+a_{n+1}}{2})\right| > |a_n|^c \, \left|\frac{a_{n+1}-a_n}{2a_n}\right|^{\mu_n} \, \left|\frac{a_{n+1}-a_n}{2a_{n+1}}\right|^{\mu_{n+1}}$$

$$> \frac{1}{16} \, |a_n|^c (a_{n+1}-a_n)^{\mu_n+\mu_{n+1}} \, \frac{1}{a_n^2 \cdot a_{n+1}^2}$$

Thus if $|a_{n+1}| \leq 2|a_n|$

$$\left|f(\frac{a_n+a_{n+1}}{2})\right| > \frac{1}{64} \, |a_n|^{12} \, (a_{n+1}-a_n)^4 \quad \text{since } c > 16 \text{ while if}$$

$$|a_{n+1}| > 2|a_n|$$

$$\left|f(\frac{a_n+a_{n+1}}{2})\right| > \frac{|a_n|^{14}}{64} \, (a_{n+1}-a_n)^2 \quad \text{or} \quad \frac{|a_n|^{15}}{32} \, (a_{n+1}-a_n)$$

depending on whether $\mu_n = 2$ or 1.

Thus in all three cases (using $a_n > 3$)

$$\left|f(\frac{a_n+a_{n+1}}{2})\right| > a_n^{10} \, (a_{n+1}-a_n) > 1 \quad \text{for all large enough n.}$$

Since for large n, $a_{n+1}-a_n > \lambda_n \exp [(n+1)^{1/p} - n^{1/p}]$

$$> \exp n^{1/p} \, [\frac{1}{p} \, n^{\frac{1}{p-1}}] > 1.$$

We note that $|a_n-a_{n-1}| > 1$ and $|f(r)| > 1$.

Thus one of $f(r)$, $\quad f(\frac{a_n+a_{n+1}}{2})$

is positive and greater than 1, since the only zero in

$$(r, \ \frac{a_n+a_{n+1}}{2})$$

is $a_n$ and $f(z)$ is positive in at least one of $(r, a_n)$, $(a_n, a_{n+1})$. But the "ones" are at $a_k$ so we cannot have any in $(a_n \cdot a_{n+1})$. Hence, there must be a "one" in $(r, a_n)$. Now in $r \leq |z| < |a_n|$

$$|f(z)| = \left| Q(z)(1 - z/a_n)^{\mu_n} \prod_{m=n+1}^{\infty} (1 - z/a_m)^{\mu_m} \right|$$

$$> |Q(z)| \, (1 - z/a_n)^{\mu_n} |P_n|^2$$

$$\geq \frac{r^{3c}}{|a_n|^{2c}} (1 - z/a_n)^{\mu_n}$$

If for some $s, r < s < a_n$, $f(s) = 1$

Then

$$1 > \frac{r^{3c}}{|a_n|^{2c}} (1 - s/a_n)^{\mu_n}$$

Thus $(1 - s/a_n)^{\mu_n} < \dfrac{|a_n|^{2c}}{r^{3c}} < \dfrac{a_n^{2c}}{\frac{9}{4}^c a_n} = \dfrac{1}{(a_n)^{c/4}}$

$$< \frac{1}{|a_n|^4} \quad \text{since } c > 16 .$$

Hence $\left| \dfrac{a_n - s}{a_n} \right|^{\mu_n} < \dfrac{1}{|a_n|^4}$ , $\mu_n = 1$, or $2$

Thus $|a_n - s| < \dfrac{1}{|a_n|^3}$ or $\dfrac{1}{a_n}$

so $|a_n - s| < \dfrac{1}{|a_n|} < 1$

However $|a_n - s| < \dfrac{1}{a_n} < 1$ contradicts the fact that $|a_n - a_{n-1}| > 1$.

Hence $E = \{a_n\}$ is a Picard set.


APPLICATION OF AN EXCEPTIONAL SET RESULT TO FACTORIZATION OF AN ENTIRE FUNCTION:

We shall look at the question of expressing an entire function $F(z)$ in the form
$$F(z) = f(g(z)) \quad \text{where } f \text{ and } g \text{ are entire.}$$
Such an expression is a factorization of $F(z)$. We shall call $F$ pseudoprime [prime] if every such factorization of $F$ implies that at least one factor is a polynomial [linear function].

A result of Polya [10] useful in our discussion states, that if f and g are transcendental entire functions such that f(g) is of finite order, then f is of zero order.

As an application of an exceptional set result to factorization theory, we give the proof of the following result of Liverpool [7].

THEOREM C.

Let E be the union of a set of discs $D_n$ of centre $z_n$ and radius $\rho_n$; where

$$\left| \frac{z_{n+1}}{z_n} \right| > q > 1, \; n = 1,2, \ldots \tag{2}$$

and $\rho_n \to 0$ as $n \to \infty$, such that E is a Picard set. Let F(z) be a transcendental entire function of finite order whose zeros all lie in E. Then F is pseudoprime and if F has an infinity of zeros at least one of which is simple, then F is prime.

REMARK

Toppila [11] has shown that the condition $\left| \frac{z_{n+1}}{z_n} \right| > q > 1 \; n=1,2,\ldots$ and $|z_n|^n = 0\log(1/\rho_n)$ together imply E is a Picard set. Liverpool has shown that there is a constant $k = k(q)$ such that the first condition together with

$$|z_n|^k = 0 \; \log(1/\rho_n) \; \text{imply that}$$

E is a Picard set, and several authors have worked on finding the best possible value of k.

PROOF :

We consider an entire function F(z) of finite order whose zeros lie within E. We suppose that F(z) = f(g(z)), where f and g are transcendental. Then, by Polya, f(t) = 0 has infinitely many solutions $t_1$, $t_2$, .... Since g is not a polynomial, $g(z) = t_1$ has solutions outside the set E for some $t_1$ as E defines a Picard set for transcendental entire functions. Hence, there is a $z^1$ which is not in E such that

$$F(z^1) = f(g(z^1)) = f(t_1) = 0,$$

against our hypothesis. Hence, F is at least pseudoprime. We now go on to show that F is indeed prime if it has an infinity of zeros at least one of which is simple.

Suppose F(z) as defined can be factorized such that F(z) = f(g(z)), with f(z) a polynomial. F(z) = f(g(z)) = 0 at an infinity of points $\xi_n$ in E. For large $\xi_n$, we have $t_n \; ( = g(\xi_n))$ large and also any other solution $\xi'$ of

$$t_n = g(\xi^1)$$

satisfies

$$F(\xi^1) = f(g(\xi^1)) = f(t_n) = 0.$$

If $g$ is a polynomial of degree $m > 1$, then as $t_n \to \infty$, $g(\xi^1) = t_n$ has m solutions (asymptotically) equally spaced on a circle viz $\xi^1 = n\lambda|t_n|^{1/m} + 0(1)$ ($\lambda$ constant), where n runs through the m-th roots of unity. However, each disc $D_n$ of E lies in an annulus

$$A_n : q^{-1/2}|z_n| < |z| < q^{1/2}|z_n|$$

and the $A_n$ are disjoint. Since $\rho_n \to 0$, each disc $D_n$, and therefore the annulus $A_n$, can contain at most one of the roots of $g(\xi^1) = t_n$ for large $t_n$. This gives a contradiction.

It remains to consider the case when f is a polynomial and g transcendental. Suppose $f = 0$ at $\alpha$ and $\beta$. Then g takes either $\alpha$ or $\beta$ outside E and so F has zeros outside E. This is impossible, so $f = 0$ at only one point, say $\alpha$, and has the form $f = \mu(w-\alpha)^k$, ($\mu \neq 0$). If $k > 1$ then $F = \mu(g-\alpha)^k$ and every zero of F has multiplicity at least k. Since F has at least one simple zero we conclude that $k = 1$ and F is prime.

REMARK

We can obtain examples of any finite order to which the theorem applies, for example, by taking canonical products. The conditions on the existence of an infinity of zeros, some simple, cannot be dropped as shown by the examples

$$p(z)e^{p(z)},$$

where $p(z)$ is a polynomial with zeros in E, and

$$(g(z))^k, \quad k > 1,$$

where g is a function of finite order with zeros in E, these functions are pseudoprime but not prime.

ACKNOWLEDGEMENT

This Research effort was supported by a grant from the University of Jos, Nigeria.

## REFERENCES

[1]    I.N. Baker, Zusammensetzungen ganzer funktionen Math Z, (69) 1958.

[2]    I.N. Baker and L.S.O. Liverpool, Picard sets of entire functions, Mat Zeit 126 (1972).

[3]    I.N. Baker and L.S.O. Liverpool, Further results on Picard sets of entire functions.  Proc. London Math. Soc. 26 (1973) 32-98.

[4]    R.P. Boas, Entire functions, New York (1954).

[5]    Holland, Introduction to the theory of entire functions, Academic Press (1973).

[6]    O. Lehto, A generalization of Picards theorem, Ark Math. 3 (1958) 495-500.

[7]    L.S.O. Liverpool, Some remarks on factorization of entire functions, J. London Math. Soc. (2) 7 (1973).

[8]    L.S.O. Liverpool and U. Umar, Some remarks on point Picard sets for entire functions.  Publications de L'institut Mathematiques 1982, 32 (46).

[9]    K. Matsumoto, Remarks on Lehto's paper (A generalization of Picards theorem) Proc. Japan Acad. 38 (1962) 636-840.

[10]   G. Polya, On an integral function of an integral function, J. London Math. Soc. 2 (1926) 12-15.

[11]   S. Toppila, Some remarks on the value distribution of entire functions, Ann. Acad. Sc. Fenn. A.I. 421 (1968).

DEPARTMENT OF MATHEMATICS
UNIVERSITY OF JOS
JOS, NIGERIA

Contemporary Mathematics
Volume 25, 1983

# REMARKS ON HOLOMORPHIC MAPPINGS

by

## Seiki Mori

**Introduction.** H. Cartan [5] and L. Ahlfors [1] extended the Nevanlinna theory to a non-degenerate holomorphic curve in $P^m(C)$ and they obtained a defect relation $\Sigma\delta(H_j) \leq m+1$, where $\{H_j\}$ are hyperplanes in $P^m(C)$ in general position and $\delta$ denotes the Nevanlinna defect. Also, Carlson-Griffiths-King theory [4] and [8] showed a defect relation for holomorphic mappings $f\colon C^n \longrightarrow V^m$ where $V^m$ is a smooth projective algebraic variety of dimension m ($\leq n$). However it is an open problem to determine a good upper bound for $\Sigma\delta(D_j)$ for a holomorphic mapping $f\colon C^n \longrightarrow P^m(C)$ (m>n) and any collection $\{D_j\}$ of hypersurfaces of degree p > 1 in $P^m(C)$ with normal crossings. B. Shiffman [9] and A. Biancofiore [3] showed that good upper bound for $\Sigma\delta(D_j)$ for holomorphic mappings belonging to special classes.

On the other hand, Nevanlinna asked the question whether the defect relation $\sum_{j=1}^{q} \delta(\phi_j,f) \leq 2$ is valid for mermorphic function $f(z)$ and distinct meromorphic functions $\phi_j(z)$ with $T_{\phi_j}(r) = o(T_f(r))$, $(r\to\infty)$ $(j=1,\ldots,q)$. (This is considered as the defect relation for moving divisors.) This problem was showed to the case of q=3 by Nevanlinna, and the case that $\delta(\infty,f)=1$ and q is any positive integer by C.T. Chuang [6]. Recently, Shiffman [10] generalized the result of Chuang [6] to the case of $C^n$ under some conditions.

In this note, we show an upper bound for $\Sigma\delta(D_j)$ for a holomorphic mapping $f\colon C \longrightarrow P^mC$ and any collection of hypersurfaces $\{D_j\}$ in $P^mC$ of degree d with some crossings conditions using a method by Cowen and Griffiths [7]. Furthermore, we give some applications for moving divisors.

1. **Notation and terminology.** Let $z=(z_1,\ldots,z_n)$ be the natural coordinate system in $C^n$. We set $||z||^2 = \sum_{j=1}^{n}|z_j|^2$, $B(r)=\{z\epsilon C^n|\ ||z|| <r\}$, $\partial B(r)=\{z\epsilon C^n|\ ||z||=r\}$, $\psi=dd^c\log ||z||^2$ $\psi_k=\psi\wedge\ldots\wedge\psi$ (k-times) and $\sigma=d^c\log ||z||^2 \wedge\psi_{n-1}$ where $d^c= \frac{\sqrt{-1}}{4\pi}(\bar\partial-\partial)$. The counting function $N(\hat{D},r)$ of a divisor $\hat{D}$ in $C^n$ is defined as usually by

$$N(\hat{D},r) = \int_1^r \frac{dt}{t} \int_{B(t) \cap \hat{D}} \psi_{n-1}.$$

Let $P^m(C)$ be an m-dimensional complex projective space, $L=[H]^d$ the line bundle of d-th tensor power of the hyperplane bundle $[H]$ over $P^m(C)$ and $|L|$ the complete linear system of effective divisors given by the zeros of holomorphic sections of $L \longrightarrow P^m(C)$. Let $\omega = dd^c \log \|w\|^{2d}$ ($w \varepsilon C^{m+1}-\{0\}$) be the curvature form of the canonical hermitian metric in $L=[H]^d$. For a holomorphic mapping $f: C^n \longrightarrow P^m(C)$ and a divisor $D \varepsilon |L|$, we define the characteristic function $T_f(L,r)$ of f, the counting function $N(D,r)$ and the defect $\delta(D,f)$ of D by

$$T_f(L,r)= \int_1^r \frac{dt}{t} \int_{B(t)} f^* \omega \wedge \psi_{n-1},$$

$$N(D,r)=N(f^* D,r)$$

and

$$\delta(D,f)=1-\limsup_{r\to\infty} (N(D,r)/T_f(L,r)),$$

where $f^* X$ denotes the pullback of X under f. (See [4],[7].) In particular, we use notation $T_f(r)$ instead of $T_f([H],r)$ for simplicity. We shall use notation " $A(r) \leq B(r)$ //" to means that the stated inequality holds outside an exceptional set E such that $\int_E \frac{dt}{t} < \infty$. ([7, p. ]04]).

We note that for a holomorphic mapping $f: C^n \longrightarrow P^m(C)$ with a reduced representation $f=(f_0,\ldots,f_m)$, its characteristic function $T_f([H]^d,r)$ is written in the form

$$T_f([H]^d,r) = \int_{\partial B(r)} \log f^* h \, \sigma + 0(1)$$

where h is a metric in $[H]^d$.

2. Defect relation for holomorphic curves. Using the methods in Cowen-Griffiths [7], we can obtain the estimate on sum of defects as follows:

Proposition 2.1. Let $f: C \longrightarrow P^m C$ be a non-degenerate (in the sense that $f(C) \not\subset D$ for any $D \varepsilon |[H]^d|$) holomorphic curve and $\{D_j\}_{j=1}^q$ a collection of hypersurfaces in $P^m C$ of degree d such that $\bigcap_{k=1}^{h+1} \text{supp} \, D_{j_k} = \phi$ for some fixed positive integer h. Then

$$\sum_{j=1}^q \delta(D_j,f) < h \, (N+1), \text{ where } N = \binom{m+d}{m} - 1.$$

The proof is done using the method due to Cowen-Griffiths [7].

Now we consider the Veronese mapping $v_d : P^m C \longrightarrow P^N C$ of degree d. Then a hypersurface $D_j (\subset P^m C)$ of degree d is mapped to a hyperplane section $A_j \cap v_d (P^m C)$ for some hyperplane $A_j \subset P^N C$ and $\bigcap_{k=1}^{h+1} A_{j_k} \cap v_d (P^m C) = \phi$.

We use notations $|B,C| = |\langle B,C \rangle|$ for the norm of interior product $\langle B,C \rangle$ of a k-vector B and an h-vector C in $C^{m+1}$.

$$\psi_i (A_\nu) = |F(z) \wedge F''(z) \wedge \cdots \wedge F^{(i)}(z), A_\nu|^2 / |F(z) \wedge F'(z) \wedge \cdots \wedge F^{(i)}(z)|^2$$

for $F = V_d \circ f$ and $\mu > 0$ as in [7, p. 104 and pp. 125-126]. Let $T_k(r) \equiv T_{F,k}(r)$ be the order function for the k-th associated curve [7, p. 116], $N_i (A_\nu, r)$ the counting function for the points $z \varepsilon C$ where $f(C)$ has contact of order i+1 with $A_\nu$ and $N_i(r)$ the counting function for the stationary points of index i [7, pp. 132-133].

Lemma 2.2 (cf. [7, Lemma 6.13]).  Let $f, v_d$ and $A_j$ be as above.  Set $H_j \equiv A_j \cap v_d (P^m C)$ (j=1,...,q) and $F \quad v_d \circ f: C \longrightarrow P^N C$.  Then F is linear non-degenerate $\bigcap_{k=1}^{h+1} H_{j_k} = \phi$ for any $\{j_1 j_2, \ldots j_{h+1}\}\{1,2,\ldots,q\}$ and

$$\sum_{\nu=1}^{q} \frac{\psi_{i+1}(A_\nu)}{\psi_i(A_\nu) (\log \frac{\mu}{\psi_i(A_\nu)})^2} \geq c_i \prod_{\nu=1}^{q} \left( \frac{\psi_{i+1}(A_\nu)}{\psi_i(A_\nu) (\log \frac{\mu}{\psi_i(A_\nu)})^2} \right)^{1/h}$$

where $c_i$ is a constant.

Proof.  Let $Z_0, \ldots, Z_N$ be a unitary frame for F at $Z = F(z)$.  Then $\psi_i(A_\nu) = |Z_0, A_\nu|^2 + \ldots + |Z_i, A_\nu|^2$, hence $|Z_0, A_\nu|^2 = 0$ if $\psi_i(A_\nu) = 0$.  Since $Z_0 = e^{i\psi} Z/|Z|$ and $\bigcap_{k=1}^{h=1} H_{j_k} = \phi$, there are at most h-$H_j$'s intersecting to F(z) on $v_d (P^m C)$.  Thus the number of $A_\nu$ with $\psi_i(A_\nu) = 0$ is at most h (i=0,1,....,N).

While $v_d (P^m C)$ is compact and $|Z_0, A_\nu|^2$ are positive and continuous in $z \varepsilon P^N C$, so there is a positive number m > 0 such that $m \leq \psi_i(A_\nu)$ $(\leq \psi_{i+1}(A_\nu) \leq ||A_\nu^*|| \leq 1)$, for all but at most h of $A_\nu$'s.  Set

$$\phi_i^{\nu} \equiv \phi_i(A_{\nu}) \equiv \phi_{i+1}(A_{\nu})/\phi_i(A_{\nu}) \left(\log \frac{\mu}{\phi_i(A_{\nu})}\right)^2 .$$

Then there is a constant $K > 0$ such that $\phi_i(A_{\nu}) \leqq K$ for all but at most h of $A_{\nu}$'s and $\phi_i(A_{\nu}) > K$ at most for $\nu = 1, \ldots, p \leqq h$ (if necessary, we rearrange indicies). Therefore we have

$$\sum_{\nu=1}^{q} \phi_i^{\nu} \geqq \sum_{\nu=1}^{p} \phi_i^{\nu} \geqq c_i' \prod_{\nu=1}^{p} \{\phi_i^{\nu}\}^{1/p} \geqq c_i \prod_{\nu}^{q} \{\phi^{\nu}\}^{1/h} \quad \text{at } z.$$

We define $\omega_i \equiv c_i \left\{ \prod_{\nu=1}^{q} \left( \dfrac{\psi_{i+1}(A_{\nu})}{\psi_i(A) \left(\log \frac{\mu}{\psi_i(A_{\nu})}\right)^2} \right)^{1/h} \right\} \Omega_i$

where $\Omega_i \equiv dd^c \log |z_0 \wedge \ldots \wedge z_i|^2$. Then, we have

Lemma 2.3.   (cf.[7, Lemma 6.4]).

$$h \cdot \sum_{\nu=0}^{N-1} \text{Ric } \omega_i \geqq (q - h)\Omega_0 - h\Omega_{N-1} + 2 \sum_{i=0}^{N-1} \left\{ \sum_{\nu} \left( \frac{\psi_{i+1}(A_{\nu})}{\psi_i(A_{\nu}) \left(\log \frac{\mu}{\psi_i(A_{\nu})}\right)^2} \right)^{1/h} \Omega_i \right\}$$

$$- \varepsilon \sum_{i=0}^{N-1} \Omega_i .$$

For any $\varepsilon > 0$, where Ric $\Omega_i$ denotes the Ricci form of $\Omega_k$.

Proof.   Note that Ric $\Omega_k = dd^c \log h_k = \Omega_{k+1} - 2\Omega_k + \Omega_{k-1}$,

where $\Omega_k = \dfrac{\sqrt{-1}}{2\pi} h_k(z) \, dz \wedge d\bar{z}$. Hence we see

$$\sum_{i=0}^{N-1} \text{Ric } \Omega_i = \sum_{i=0}^{N-1} (\Omega_{i+1} - 2\Omega_i + \Omega_{i-1}) = -(\Omega_0 + \Omega_{N-1}) .$$

On the other hand, from [7, (6.11)]

$$dd^c \log \left( \frac{1}{\left(\log \frac{\mu}{\phi_i(A_{\nu})}\right)^2} \right) \geqq -\varepsilon\Omega_j + \frac{2\phi_{i+1}(A_{\nu}) \, \Omega_i}{\phi_i(A_{\nu}) \left(\log \frac{\mu}{\phi_i(A_{\nu})}\right)^2}$$

for sufficiently large $\mu$. Therefore we have

$$\sum_{i=0}^{N-1} dd^c \log \left\{ \prod_{\nu=1}^{q} \frac{\phi_{i+1}(A_{\nu})}{\phi_i(A_{\nu}) \left(\log \frac{\mu}{\phi_i(A_{\nu})}\right)^2} \right\}$$

$$= - \sum_{\nu=1}^{q} dd^{c} \log \phi_{0}(A_{\nu}) + \sum_{i=0}^{N-1} \sum_{\nu} dd^{c} \log \left\{ \frac{1}{(\log \frac{\mu}{\phi_{i}(A_{\nu})})^{2}} \right\}$$

$$\geq q\, \Omega_{0} + 2 \sum_{i=0}^{N-1} \left\{ \sum_{\nu} \frac{\phi_{i+1}(A_{\nu})}{\phi_{i}(A_{\nu})(\log \frac{\mu}{\phi_{i}(A_{\nu})})^{2}} \, \Omega_{i} \right\} - \epsilon' \sum_{i=0}^{N-1} \Omega_{i}.$$

Thus we obtain

$$h \sum_{i=1}^{N-1} \mathrm{Ric}\, \omega_{i} = q\Omega_{0} + h \sum_{i=0}^{N-1} \mathrm{Ric}\, \Omega_{i} + 2 \sum_{i=0}^{N-1} \left\{ \sum_{\nu} \frac{\phi_{i+1}(A_{\nu})\, \Omega_{i}}{\phi_{i}(A_{\nu})(\log \frac{\mu}{\phi_{i}(A_{\nu})})^{2}} \right\}$$

$$- \epsilon \sum_{i=0}^{N-1} \Omega_{i}$$

$$= (q-h)\, \Omega_{0} - h\, \Omega_{N-1} + 2 \sum_{i=0}^{N-1} \omega_{i} - \epsilon \left( \sum_{i=0}^{N-1} \Omega_{i} \right).$$

Next we see the following.

Lemma 2.4. (cf. [7, Lemma 6.5]).

$$\sum_{\nu=1}^{q} N(A,r) \geq (q - h(N+1) - \epsilon)\, T_{0}(r) + \mathrm{const.} \qquad //.$$

Proof.  Using similar method to [7], we have

$$(2.1) \quad \frac{h}{4\pi} \int_{|z|=r} \log h_{i}\, d\theta = hN_{i}(r) + \sum_{\nu=1}^{q} \{ N_{i+1}(A_{\nu},r) - N_{i}(A_{\nu},r) \}$$

$$+ h \int_{0}^{r} \left( \int_{|z|<\delta} \mathrm{Ric}\, \omega_{i} \right) \frac{d\rho}{\rho} + c_{i}, \qquad (z=re^{i\theta}).$$

From [7, (2.6) and (2.11)],

$$(2.2) \quad \frac{h}{4\pi} \int_{|z|=r} \log h_{i}\, d\theta \leq \log \frac{1}{2\pi} \int_{|z|=r} h_{i}\, d\theta = c' \log (T_{\omega_{i}}(r)) //,$$

where $T_{\omega_{i}}(r) \equiv \int_{0}^{r} \frac{d\rho}{\rho} \int_{|z|<r} \omega_{i}$. Using [7, (4.22) $(\ell+1)T_{k}(r) \leq (k+1)T_{\ell}(r) + \epsilon(T_{0}(r))$ for $\ell<k$],

$$(2.3) \quad \epsilon \left( \sum_{i=0}^{N-1} T_{i}(r) \right) \leq c'' T_{0}(r) \qquad //.$$

Summing (2.1) for i = 1 to N-1 and using Lemma 2.3, (2.2) and (2.3), we obtain

$$\sum_{\nu=1}^{q} N(A_{\nu},r) \geqq \sum_{i=0}^{N-1} \sum_{\nu=1}^{q} \{N_i(A_{\nu},r) - N_{i+1}(A_{\nu},r)\}$$

$$\geqq h \sum_{i=0}^{N-1} N_i(r) - \varepsilon T_0(r) + \int_0^r \left\{ \int_{|z|<\rho} (q-h)\, \Omega_0 - h\, \Omega_{N-1} \right.$$

$$\left. + 2 \sum_{i=1}^{N-1} \hat{\omega}_i - \varepsilon (\sum_{i=1}^{N-1} \Omega_i) \right\} \frac{d\rho}{\rho} + c$$

$$\geqq (q-h)\, T_0(r) - h\, T_{N-1}(r) + 2 \sum_{i=0}^{N-1} T_{\omega_i}(r) - \varepsilon \sum_{i=0}^{N-1} T_i(r) + c$$

$$\geqq (q - h - h\,N)\, T_0(r) - \varepsilon\, T_0(r) + c$$

$$\geqq (q - h\,(N+1) - \varepsilon)\, T_0(r) + c \quad //.$$

Proof of Proposition 2.1. We note that $N(f^*D_{\nu},r) = N(F^*A_{\nu},r)$ $(\nu=1,\ldots,q)$ and $T_f(r) = d.T_F(r) + 0(1)$. Hence we have

$$\sum_{\nu=1}^{q} \delta(D_{\nu},f) = \sum_{\nu=1}^{q} \delta(A_{\nu},F) = \sum_{\nu=1}^{q} (1 - \limsup_{r\to\infty} (N(A_{\nu},r)/T_0(r))$$

$$\leqq q - \limsup \sum_{\nu=1}^{q} (N(A_{\nu},r)/T_0(r)) \leqq h(N+1) + \varepsilon$$

for any $\varepsilon > 0$. Thus we obtain

$$\sum_{\nu=1}^{q} \delta(D_{\nu},f) \leqq h(N+1). \qquad\qquad\qquad (Q.E.D.)$$

Let $V \equiv V^m$ be a projective algebraic manifold of dimension, m, $L \longrightarrow V$ a very ample line bundle over V with dim $\Gamma(V,\sigma(L)) = N+1$.

Remark $1^{\circ}$. Let $f : C \longrightarrow V$ be a holomorphic mapping and $s = (s_0 : \ldots : s_N) : V \hookrightarrow P^N C$ an embedding given by basis of sections in $\Gamma(V,\mathcal{O}(L))$. For a divisor $D \varepsilon |L|$, there is a hyperplane $H \subset P^N C$ such that $\tilde{D} \equiv s(D) = H \cap s(V)$. Hence if $\{D_j\}_{j=1}^{q} \varepsilon |L|$ is any collection of divisors on V such that $\bigcap_{k=1}^{h+1} \text{supp } D_{j_k} = \phi$, we see

$$\sum_{j=1}^{q} \delta(D_j, f) \leq s\,(N+1).$$

$2^{\circ}$. By Siegel's theorem, dim $\Gamma(V, \sigma(L^d)) \leq 0(d^m)$ and $\binom{m+d}{m} \ll d^m$ ($m \ll 1$).

$3^{\circ}$. If $\{D_j\}_{j=1}^{q}$ has normal crossings, then $\bigcap_{K=1}^{m+1} \mathrm{supp}\, D_{j_k} = \phi$, so we can

take $s = m$.

3. **Moving divisor.**

Definition. We shall say that $D^M \equiv D(z) : \mathbb{C}^n \longrightarrow |[H]^d|$ is a moving divisor

of degree d in $P^m(\mathbb{C})$ if for any fixed $z \in \mathbb{C}^n$ it is a hypersurface of degree d

(counted according to multiplicity) in $P^m(\mathbb{C})$ which is written in the form $D(z)$

$= \{w \in \mathbb{C}^{m+1} - \{0\} \mid P(z,w) = 0\}$ where $P(z,w)$ is a homogeneous polynomial of degree d

in w whose coefficients are entire functions in z. We set $N(D^M, r) = N(D_f^M, r)$

where $D_f^M = |z \in \mathbb{C}^n | P(z, f(z)) = 0\}$. For a moving divisor $D^M$ of degree d and a

holomorphic mapping $f : \mathbb{C}^n \longrightarrow P^m(\mathbb{C})$, we define the defect $\delta(D^M, f)$ by

$$\delta(D^M, f) = 1 - \limsup_{r \to \infty} (N(D^M, r) / T_f([H]^d, r)).$$

First of all, we shall treat the case of $n = 1$ and $d \geq 1$. We shall show a

defect relation for moving divisors using Proposition 2.1.

Let $\{D_j^M\}_{j=1}^{q}$ be a collection of moving divisors of degree d in $P^m(\mathbb{C})$.

We write $D_j^M \equiv D_j(z) = \{w \in \mathbb{C}^{m+1} - 0\} |P_j(z,w) = 0\}$ and $P_j(z,w) = \sum_{|k|=d} a_k^j(z) \cdot m_k(w)$

where $m_k(w)$ are monomials of degree d in w and $a_k^j(z)$ are entire functions in

$z \in \mathbb{C}$. Here k is a multi-indices of length $|k|$. Suppose that there is a system

of $(h+1)$ entire functions $\zeta(z) = (\zeta_0(z), \ldots, \,_h(z))$ without common zeroes such

that each entire function $a_k^j(z)$ is written in the form $a_k^j(z) = Q_k^j(\zeta(z))$ for any

j and k. Here $Q_k^j(\zeta(z))$ are homogeneous polynomials of degree d' in $\zeta_0, \ldots, \zeta_h$.

For example if $a_k^j$ are polynomials of degree d' in z, we can take

$\zeta(z) = (1, z)$.

We say that a holomorphic curve $f : \mathbb{C} \longrightarrow P^m(\mathbb{C})$ is non-degenerate with

respect to $\mathbb{C}[\zeta]$ if for any moving divisor $D^M \equiv D(z)$ of degree d in $P^m(\mathbb{C})$ whose

defining equation has coefficients in the polynomial ring $\mathbb{C}[\zeta]$ of the system

$\zeta=(\zeta_0,\ldots,\zeta_h)$ over C, there is a point $z_0 \epsilon C$ satisfying $f(z_0) \notin D(z_0)$.

Now $\zeta(z)$ can be considered as a holomorphic curve of C into $P^h(C)$ and $D_j^M$ moving divisors $D_j^M \equiv D_j(\zeta): C^{h+1}-\{0\} \longrightarrow |[H]^d|$.

Theorem 3.1. Let $f: C \longrightarrow P^m(C)$ be a holomorphic curve which is non-degenerate with respect to $C[\zeta]$, $\{D_j^M\}_{j=1}^q: C \longrightarrow |[H]^d|$ a collection of moving divisors of degree d in $P^m(C)$ as above satisfying $\bigcap_{i=1}^{s+1} \operatorname{supp} D_{j_i}^m = \emptyset$ for any $(j_1,\ldots,j_{s+1}) \subset (1,\ldots,q)$ and any fixed $\zeta \epsilon C^{h+1}-\{0\}$. Suppose that $T_\zeta(r)=o(T_f(r))$ $(r\to\infty)$. Then

$$\sum_{j=1}^q \delta(D_j^M,f) \leqq s\cdot(R+1),$$

where $R = \binom{h+d'}{h}\binom{m+d}{m}-1$.

Proof. For a holomorphic curve $f: C \longrightarrow P^m C$, we consider the holomorphic curve curve $\tilde{f}: C \longrightarrow P^h(C) \times P^m(C)$ given by $\tilde{f}(z)=(\zeta(z),f(z))$, and further we consider the Veronese mappings $v=(v_0,\ldots,v_R): P^h(C) \times P^m(C) \longrightarrow P^R(C)$ given by $v_j=P_{d'}^j(\zeta)Q_d^j(w)$ where $P_{d'}^j(\zeta)$ are monomials of degree d' in $\zeta$ and $Q_d^j(w)$ monomials of degree d in w. Set $F=(F_0,\ldots F_R) \equiv v\circ f$. Let $[H_1]$, $[H_2]$ and $[H]$ be the hyperplane bundles over $P^h(C)$, $P^m(C)$ and $P^R(C)$, respectively. Then $g=\sum_{j=0}^R |\eta_j|^2$ $(\eta=(\eta_0,\ldots,\eta_R) \epsilon C^{R+1}-\{0\})$ is a metric in $[H]$, so $v^*g$ is a metric in the line bundle $[H_1]^{d'} \otimes [H_2]^d$. On the other hand $(\sum_{j=0}^h |\zeta_j|^{2d'})$, $(\sum_{j=0}^m |w_j|^{2d})$ and $(\sum_{j=0}^h |\zeta_j|^{2d'})(\sum_{j=0}^m |w_j|^{2d})$ are also metrices in $[H_1]^{d'}$, $[H_2]^d$ and $[H_1]^{d'} \otimes [H_2]^d$, respectively. Hence we have

$$T_F([H],r) = \int_0^{2\pi} \log \sum_{j=0}^R |F_j(re^{i\theta})|\, d\theta + 0(1)$$

$$= \int_0^{2\pi} \log \sum_{j=0}^R |P_{d'}^j(\zeta(re^{i\theta}))Q_d^j(f(re^{i\theta}))|\, d\theta + 0(1)$$

$$= \int_0^{2\pi} \log(\sum_{j=0}^h |\zeta_j(re^{i\theta})|^{d'})(\sum_{j=0}^m |f_j(re^{i\theta})|^d)d\theta + 0(1)$$

$$= \int_0^{2\pi} \log \sum_{j=0}^{h} |\zeta_j(re^{i\theta})|^{d'} d\theta + \int_0^{2\pi} \log \sum_{j=0}^{m} |f(re^{i\theta})|^{d} d\theta + 0(1)$$

$$= d'T_\zeta(r) + dT_f(r) + 0(1) = (1 + o(1))dT_f(r).$$

Take $\tilde{D}_j \equiv \{(\zeta,w) \in (C^{h+1}-\{0\}) \times (C^{m+1}-\{0\}) \mid P_j(\zeta,w) = 0\}$ where $P_j(\zeta,w)$ are homo-geneous polynomials of degree d in w and of degree d' in $\zeta$. Then the support of divisors $\tilde{D}_j$ (j=1,...,q) are mapped to hyperplane sections $H_j$ in $P^R(C)$ under

v and $\bigcap_{k=1}^{s+1} H_{j_k} = \emptyset$ for any $(j_1,...j_{s+1}) \subset (1,...,q)$ on the compact set $v(P^h(C) \times$

$P^m(C))$. Clearly we see $F^*H_j = \tilde{f}^*\tilde{D}_j = D^M_{j},f$. Since f is non-degenerate with

respect to $C[\zeta]$,F is not linearly degenerate. Therefore we obtain from

Proposition 2.1

$$\sum_{j=1}^{q} \delta(D^M_j,f) \leq s \cdot (R+1).$$
(Q.E.D.)

Corollary 3.2. Let $\{D^M_j\}^q_{j=1}$ be a collection of moving divisors as above.

If $f(z) \cap \bigcup_{j=1}^{(R+1)s}$ supp $D^M_j = \emptyset$ for any $z\in C$ or if $\sum_{j=1}^{q} \delta(D^M_j,f) > (R+1)s$, then f is

degenerate with respect to $C[\zeta]$.

In the case of $C^n$ $(n\geq 1)$, let $\zeta_0,...,\zeta_h$ be a basis of $\{a^j_k(z)\}$. Suppose

that $\zeta_0,...,\zeta_h$ has no common factors. Then using [2], we can easily show the

following:

Theorem 3.3. Let $f: C^n \longrightarrow P^m(C)$ be a holomorphic mapping and

$\{D^M_j\}^{h+m+2}_{j=1}$ a collection of moving divisors on $P^m(C)$ such that $\bigcap_{k=1}^{h+m+1}$ supp $D^M_{j_k} = \emptyset$

for any fixed $\zeta \in C^{h+1}-\{0\}$ and any $(j_1,...,j_{h+m+1}) \subset (1,...,h+m+2)$. If

$\zeta(D^M_j,f)=1$ for j=1,...,h+m+2, then f is degenerate with respect to $C[\zeta]$.

Proof. For the holomorphic $f: C^n \longrightarrow P^m(C)$, we consider the holomorphic

mapping $\tilde{f}: C^n \longrightarrow P^h(C) \times P^m(C)$ and divisors $\tilde{D}_j \equiv \{(\zeta,w) \mid P_j(\zeta,w) = 0\}$ (j=1,...,

h+m+2) on $P^h(C) \times P^m(C)$. Then $\bigcap_{k=1}^{h+m+1}$ supp $\tilde{D}_{j_k} = \emptyset$ on $P^h(C) \times P^m(C)$, $D^M_{j,f} = \tilde{f}^*\tilde{D}_j$ and

$(1+o(1))T_{\tilde{f}}([H_2]^d,r) = T_{\tilde{f}}([H_1]^{d'} \otimes [H_2]^d,r)$. Hence from [2], we see $\tilde{f}$ is

algebraically degenerate. Therefore f is degenerate with respect to $C[\zeta]$.

In particular, if $a^j_k$ are rational functions, we can take h+m+2=n+m+2.

Finally we shall show a generalization of a result of Nevanlinna for

moving hyperplanes

S. MORI

Theorem 3.4.  Let $a_j = (a_0^j(z), \ldots, a_m^j(z)): C^n \longrightarrow P^m C$ be holomorphic mappings such that $a_j = a_j(z)$ $(j=0,\ldots,m+1)$ are in general position for almost all $z \in C^n$ in the sense that $\det\{(a_k^{j_i}(z))_{0 \le i,k \le m}\} \not\equiv 0$, for any $(j_0,\ldots,j_m) \subset (0,\ldots,m+1)$. Let $H_j^M \equiv H_j(z) = \{w = (w_0,\ldots,w_m) \in C^{m+1} - \{0\} \mid \sum_{k=0}^{m} a_k^j(z) w_k = 0\}$, $(j=0,\ldots,m+1)$. If $f: C^n \longrightarrow P^m(C)$ is a holomorphic mapping such that $T_{a_j}(r) = o(T_f(r))$, $(r \to \infty)$ for $j=0,\ldots,m+1$, and suppose that $f$ is non-degenerate with respect to $C\{a_j\}$. Then

$$\sum_{j=0}^{m+1} \delta(H_j^M, f) \le m+1.$$

Here $f$ is non-degenerate with respect to $C\{a_j\}$ means that for any moving hyperplane $H^M = H(z)$ whose defining equation has coefficients consisting from homogeneous polynomials of degree $m+2$ as variables in $a_k^j$'s and of degree 1 for each system $a_j = (a_0^j, \ldots, a_m^j)$, there is a point $z_0 \in C^n$ such that $f(z_0) \not\in H(z_0)$. Note that this result is not depend on a choice of representation of holomorphic mappings $a_j$, and further the condition that $f$ is non-degenerate with respect to $C\{a_j\}$ is not dropped.

Example.  Let $f = (1, e^z, ze^z)$, $a_0 = (1,0,0)$, $a_1 = (0,1,0)$, $a_2 = (0,0,1)$ and $a_3 = (1,z,1)$ be holomorphic curves from $C$ into $P^2(C)$. Then $\{a_j\}_{j=0}^{3}$ is in general position for almost all $z \in C$ and $\delta(H_j^M, f) = 1$ for $j=0$ to 3 so $\sum_{j=0}^{3} \delta(H_j^M, f) = 4$. But $f(z) \in \{(w_0, w_1, w_2) \in C^2 - \{0\} \mid -zw_1 + w_2 = 0\}$ for any $z \in C$ and the equation $-zw_1 + w_2 = 0$ is considered as $a_1^1 a_2^2 a_3^3 a_2^4 w_1 + a_1^1 a_2^2 a_3^3 a_3^4 w_2 = 0$. Hence $f$ is degenerate with respect to $C\{a_j\}$.

Proof of Theorem 3.4.  Let $D = \det\{(a_k^j(z))_{0 \le j,k \le m}\}$ and $D_h$ is the determinant of the matrix replaced the $h$-th column vector ${}^t(a_0^h, \ldots, a_m^h)$ of the matrix $(a_k^j(z))_{0 \le j,k \le m}$ by the column vector ${}^t(a_0^{m+1}, \ldots, a_m^{m+1})$. We set $x_j \equiv D_j/D$ and $F_j(z) \equiv \sum_{i=0}^{m} a_k^j(z) f_k(z)$. Let $d(z)$ be a common factor among $x_j F_j$'s up to non-vanishing holomorphic functions. Consider the meromorphic mapping

$G \equiv (G_0, \ldots, G_m) : C^n \longrightarrow P^m(C)$ given by $G_j(z) = x_j(z)F_j(z)/d(z)$ $(j=0,\ldots,m)$.
Then $G$ is not depend on a choice of representation of $a_j$, and also $G$ is
linearly degenerate since $f$ is non-degerate with respect to $C\{a_j\}$. Set $H_j =$
$\{w=(w_0,\ldots,w_m) \varepsilon C^{m+1}-\{0\} | w_j = 0\}$ $(j=0,\ldots,m)$ and $H_{m+1} = \{w=(w_0,\ldots,w_m)\ \varepsilon C^{m+1}$
$-\{0\} \mid \sum_{i=0}^{m} w_i = 0\}$. Then we see $G^*H_j \leqq (F_j)_0 + (D_j)_0$ $(j=0,\ldots,m)$ and $G^*H_{m+1} \leqq$
$(\sum_{j=0}^{m} F_j)_0$ where $D' \leqq D''$ means that $D'' - D'$ is an effective divisor and $(g)_0$

denotes the divisor given by zeros of a holomorphic function $g$. Hence we see
$N(G^*H_j,r) \leqq N(H_j^M,r)+N((D_j)_0,r)$ $(j=0,\ldots m)$ and $N(G^*H_{m+1},r) \leqq N(H_{m+1}^M,r)$. Note
that using Hadamard inequality,

$$N((D_j)_0,r) = \int_{\partial B(r)} \log|D_j|^2 \sigma + 0(1) \leqq \int_{\partial B(r)} \log \prod_{\substack{h=0 \\ h \neq j}}^{m+1} ||a_h||^2 \sigma + 0(1)$$

$$\leqq \sum_{j=0}^{m+1} T_{a^j}(r) + 0(1) = o(T_f(r)), \quad (r \to \infty) \quad (j=0,\ldots,m),$$

and

$N((D)_0, r) = o(T_f(r)), (r \to \infty)$. Since $F_j(z) = \sum_{i=0}^{m} a_i^j(z)f_i(z)$ $(j=0,\ldots,m)$ and
$D \neq 0$, we have $f_h = E_h/D$ where $E_h$ is the determinant of the matrix replaced the
$h$-th row vector $(a_h^0,\ldots,a_h^m)$ of the matrix $(a_k^j(z))_{0 \leqq j,k \leqq m}$ by the row vector
$(F_0,\ldots,F_m)$. Thus using Hadamard inequality, we obtain $|E_k|^2||b_0||^2 \ldots ||_F^{(k)}|^2$
$\ldots || b_m ||^2$ where $||b_i||^2 \equiv \sum_{\ell=0}^{m} |a_i^\ell|^2 (\leqq \sum_{\ell=0}^{m} || a_\ell ||^2)$, Hence $\sum_{k=0}^{m} |E_k|^2 \leqq (m+1)$
$||F||^2 \sum_{\ell=0}^{m} || a_\ell ||^2)^m$. We may assume that $||a|| > 1$ on $\partial B(r)$ $(\ell=0,\ldots,m+1)$ by

exchange of representation of holomorphic mappings $a_j$ since $\partial B(r)$ is compact
and $(a_0^j,\ldots,a_m^j)$ has no common zero. So we have

$$\log \sum_{k=0}^{m} |E_k|^2 \leqq \log \{(m+1)^{m+1} ||F||^2 ( \max_{0<\ell \leqq m} ||a_\ell||^2)^m\}$$

$$\leqq m \sum_{\ell=0}^{m} \log ||a_\ell||^2 + \log ||F||^2 + (m+1) \log (m+1).$$

112    S. MORI

On the other hand, we see

$$||F||^2/|D|^2 = \sum_{k=0}^{m} |F_k|^2/|D|^2 = \sum_{k=0}^{m} |dG_k|^2/|D_k|^2 \leqq d||G||^2 \sum_{k=0}^{m} 1/|D_k|^2$$

$$= d||G||^2 (\sum_{k=0}^{m} \prod_{j\neq k}^{n} |D_j|^2)/\prod_{k=0}^{m} |D_k|^2$$

$$\sum_{k=0}^{m} \sum_{j\neq k}^{m} |D_j|^2 \leqq \sum_{k=0}^{m} (\prod_{j\neq k}^{m} ||a_0||^2 \dots ||a_{m+1}^{(k)}||^2 \dots ||a_m||^2)$$

$$\leqq (m+1)(||a_0||^2 \dots ||a_{m+1}||^2)^{m+1}.$$

Note that $(d)_0 \leqq (D)_0 + \sum_{k=0}^{m} (D_k)_0$ and hence

$$\int_{\partial B(r)} \log|d(z)|^2 \sigma = N((d)_0, r) \leqq N((D)_0, r) + \sum_{k=0}^{m} N((D_k)_0, r) = o(T_f(r)),$$

as $r \to \infty$.  Hence we have

$$T_f(r) = \int_{\partial B(r)} \log \sum_{k=0}^{m} |f_k(z)|^2 \sigma + O(1)$$

$$\leqq \int_{\partial B(r)} \log \sum_{k=0}^{m} |E_k|^2 \sigma - \int_{\partial B(r)} \log|D|^2 \sigma + O(1)$$

$$\leqq \int_{\partial B(r)} \log ||F||^2/|D|^2 \sigma + m \sum_{\ell=0}^{m} \int_{\partial B(r)} \log||a_\ell||^2 \sigma + O(1)$$

$$\leqq \int_{\partial B(r)} \log ||G||^2 \sigma + \int_{\partial B(r)} \log |d(z)|^2 \sigma$$

$$+ (m+1) \int_{\partial B(r)} \log \prod_{\ell=0}^{m+1} ||a_\ell||^2 \sigma + m \sum_{\ell=0}^{m+1} \int_{\partial B(r)} \log ||a_\ell||^2 \sigma + O(1)$$

$$= T_G(r) + O(\sum_{\ell=0}^{m+1} T_{a_\ell}(r)) \leqq T_G(r) + o(T_f(r)), \quad (r \to \infty).$$

Therefore, we obtain $(1-o(1)) T_f(r) \leqq T_G(r)$.

Thus we have by Stoll and Vitter's theorem [11], [12]

$$\sum_{j=0}^{m+1} \delta(H_j^M, f) = \sum_{j=0}^{m+1} \left(1 - \limsup_{r \to \infty} (N(H_j^M, r)/T_f(r))\right)$$

$$\leq \sum_{j=0}^{m+1} \left(1 - \limsup_{r \to \infty} (N(H_j, G)/T_G(r))\right)$$

$$= \sum_{j=0}^{m+1} \delta(H_j, G) \leq m+1. \qquad\qquad \text{(Q.E.D.)}$$

## References


[1]  L.V.Ahlfors, The theory of meromorphic curves, Acta Soc. Sci. Fenn. (Nova Ser. A) 3 (1941), 1-31.

[2]  Y. Aihara and S. Mori, Algebraic degeneracy theorem for holomorphic mappings into smooth projective algebraic varieties, Nagoya Math. J. 84 (1981), 209-218.

[3]  A. Biancofiore, A hypersurface defect relation for a class of meromorphic maps, Trans. Amer. Math. Soc. 270 (1982), 47-60.

[4]  J. Carlson and P.A. Griffiths, A defect relation for equidimensional holomorphic mappings between algebraic varieties, Ann. of Math. 95 (1972), 557-584.

[5]  H. Cartan, Sur les zeros des combinations lineaires de p fonctions holomorphes donnees, Mathematica (Cluj) 7 (1933), 5-31.

[6]  C.T. Chuang, Une generalization d'une inegalite de Nevanlinna, Sci Sinica, 13 (1964), 887-895.

[7]  M. Cowen and P.A. Griffiths, Holomorphic curves and metrices of negative curvature, J. d'Analyse Math. 29 (1976), 93-153.

[8]  P.A. Griffiths and J. King, Nevanlinna theory and holomorphic mappings between algebraic varieties, Acta. Math. 130 (1973), 145-220.

[9]  B. Shiffman, On holomorphic curves and meromorphic maps in projective space, Indiana Univ. Math. J. 28 (1979), 627-641.

[10] B. Shiffman, A general second main theorem for meromorphic functions on $C^n$, preprint.

[11] W. Stoll, Die beiden Hauptsätze der Wertverteilungs theorie bei funktionen mehrerer komplexan Veränderlichen (I) (II), Acta Math., 90 (1953), 1-115, and 92 (1954), 55-169.

[12] A. Vitter, The lemma of the logarithmic derivative in several complex variables, Duke Math. J., 44 (1977), 89-104.



DEPARTMENT OF MATHEMATICS
YAMAGATA UNIVERSITY
YAMAGATA, JAPAN

Contemporary Mathematics
Volume 25, 1983

## DEFICIENCY SUM OF HOLOMORPHIC CURVES
## OF LOWER ORDER LESS THAN ONE

by

### Kiyoshi Niino

1.   Using Baernstein's spread relation [2], Edrei [5] completely solved the deficiency problem for meromorphic functions of lower order less than one. He proved:

THEOREM A.   Let $f(z)$ be a meromorphic function of lower order $\mu(0 < \mu \leq 1)$, $\Delta(f)$ its total deficiency and $\nu(f)$ the number of its deficient values.

I.    If $0 < \mu \leq 1/2$, the inequality

$$\Delta(f) \geq 1 - \cos \pi\mu$$

is possible if and only if $\Delta(f) = 1$, $\Delta(f) = \delta_1 \leq 1$.

II.   If $1/2 < \mu \leq 1$, we always have

$$\Delta(f) \leq 2 - \sin \pi\mu. \tag{1}$$

Equality is possible in (1) if and only if $\nu(f) = 2$, $\Delta(f) = \delta_1 + \delta_2$, $\delta_1 = 1$, $\delta_2 = 1 - \sin \pi\mu$.

Petrenko - Hussain [10]  tried to extend Theorem A to holomorphic curves and obtain the following.

THEOREM B.   Let $x: C \to P_n C$ be a non-degenerate holomorphic curve of lower order $\mu$ and A a finite system of $P_n C$ in general position.

I.    If $\mu \leq 1/2$, then

$$\Sigma_{a \epsilon A} \, \delta(a) \leq n.$$

II.   If $1/2 < \mu < 1$, then for $a_j \epsilon A$,

$$\sum_{j=1}^{n+1} \delta(a_j) \leq n + 1 - \sin \pi\mu.$$

Further Krytov [8] extended (I) of Theorem A to a non-degenerate holomorphic curve of lower order $\mu(0 < \mu \leq 1/2)$.

We here note to be able to deduce from their proofs that Theorem B and the results of Krytov are also valid in case of degenerate holomorphic curves.

In this paper we mainly investigate the possibility of proving an analogous deficiency relation of (II) in Theorem A for non-degenerate or degenerate holomorphic curves of lower order $\mu$ such that $1/2 < \mu < 1$.

K. NIINO

The standard symbols of theory of holomorphic curves in the projective space are used throughout the paper (cf. Wu [12]). For meromorphic functions see Hayman [6].

2. Let $x: C \to P_n C$ be a holomorphic curve of lower order $\mu$, $\tilde{x} = (x_0, x_1, \ldots, x_n) : C \to C^{n+1} - \{0\}$ its reduced representation and $\lambda$ the maximum number of independent, linear relations with constant coefficients among the functions $x_0, x_1, \ldots, x_n$. It is clear that $0 \leq \lambda \leq n - 1$ and $\lambda$ is independent of the particular reduced representation chosen. We here call $\lambda$ the index of degeneracy of the holomorphic curve $x$. Let A be a finite system of $P_n C$ in general position. Then we here conjecture the following

$$\sum_{a \in A} \delta (a) \leq n + (\lambda + 1)(1 - \sin \pi\mu) \text{ for } 1/2 < \mu < 1. \tag{2}$$

In general we know the following Cartan's conjecture without reference to lower order [3]

$$\sum_{a \in A} \delta(a) \leq n + \lambda + 1,$$

which has not been proved yet. In special cases of $\lambda = 0$ and $\lambda = n - 1$ it was proved by Cartan [3] and Ahlfors [1].

From now on we assume that $x:C \to P_n C$ is a holomorphic curve of index $\lambda$ of degeneracy, of lower order $\mu$ such that $1/2 < \mu < 1$, that $\tilde{x} = (x_0, x_1, \ldots, x_n) : C \to C^{n+1} - \{0\}$ is its reduced representation, that A is a finite system of $P_n C$ in general position and that $T(r)$ is the characteristic (order) function of the holomorphic curve $x$.

In this paper we shall prove our conjecture (2) with additional conditions. We shall first have :

THEOREM 1. If there is a reduced representation $\tilde{x}$ of the holomorphic curve $x$ such that for $j = 1, 2, \ldots, n$

$$T(r, x_j) = o (T(r)) \quad r \to \infty ,$$

then we have

$$\sum_{a \in A} \delta (a) \leq n + (\lambda + 1)(1 - \sin \pi\mu).$$

If we put $\lambda = n - 1$ in (2), then its right-hand side is equal to $n(2 - \sin \pi\mu)$. We shall next prove

THEOREM 2. If $\lambda = n - 1$, then we have

$$\sum_{a \in A} \delta (a) \leq n(2 - \sin \pi\mu).$$

The following statement characterizes the sharpness of Theorem 1 and Theorem 2:

THEOREM 3. For any given numbers $n (n \geq 2)$, $\lambda (0 \leq \lambda \leq n - 1)$ and $\mu (1/2 < \mu < 1)$, there are a holomorphic curve $x: C \to P_n C$ of index $\lambda$ of degeneracy, of lower order $\mu$ with a reduced representation $\tilde{x} = (x_0, x_1, \ldots, x_n): C \to C^{n+1} - \{0\}$ and a system $\{a_j\}_{j=1}^{n+\lambda+1}$ of $P_n C$ in general position such that for $\nu = 1, 2, \ldots, n$,

$$T(r, x_\nu) = o(T(r)) \quad r \to \infty$$

and

$$\sum_{j=1}^{n+\lambda+1} \delta (a_j) = n + (\lambda + 1)(1 - \sin \pi\mu).$$

The author [9] and Krytov [8] extend Bearnstein's spread relation for meromorphic functions to holomorphic curves. Using this spread relation we shall have the following results. When $\lambda$ satisfies $0 \leq \lambda \leq n - 2$, we shall have:

THEOREM 4. Assume that $\lambda$ and $\mu$ satisfy $0 \leq \lambda \leq n - 2$ and $1/2 < \mu < 1$, respectively. Let $\sigma(a)$ be the spread of $a \epsilon P_n C$ such that $< x(z), a > \not\equiv 0$. If there are $a_j \epsilon P_n C$ $(j = 1, \ldots, n - 1)$ in general position such that $\sigma(a_j) = 2\pi$, then we have

$$\sum_{a \epsilon A} \delta (a) \leq n + 1 - \sin \pi\mu.$$

In the case of $\lambda = n - 1$ we shall have, under weaker conditions,

THEOREM 5. Assume that $\lambda = n - 1$, $\mu < + \infty$ and there is an $a_1 \epsilon A$ satisfying $\sigma(a_1) = 2\pi$. Then we have

$$\sum_{a \epsilon A} \delta (a) \leq n.$$

Theorem 4 and Theorem 5 are sharp in the following sense:

THEOREM 6. For any given numbers $n (n \geq 2)$, $\lambda (0 \leq \lambda \leq n - 2)$, $\mu (1/2 < \mu < 1)$ and $\epsilon (\epsilon > 0)$, there are a holomorphic curve $x: C \to P_n C$ of index $\lambda$ of degeneracy, of lower order $\mu$ and a system $\{a_j\}_{j=1}^{n+1}$ of $P_n C$ in general position such that $\sigma(a_j) = 2\pi$ $(j = 1, \ldots, n - 1)$ and

$$\sum_{j=1}^{n+1} \delta (a_j) \geq n + 1 - (1 + \epsilon) \sin \pi\mu.$$

THEOREM 7. Let $\phi(r)$ be increasing and convex in $\log r$ with $\phi(r) \neq 0 (\log r)$ $(r \to \infty)$ and $n (\geq 2)$ an integer. Then there is a holomorphic curve

$x:C \to P_n C$ of index $n - 1$ of degeneracy and a system $\{a_j\}_{j=1}^{n}$ of $P_n C$ in general position such that $T(r) \sim \phi(r)$ $(r \to \infty)$, $\sigma(a_j) = 2\pi$ $(j = 1, \ldots, n)$ and

$$\sum_{j=1}^{n} \delta(a_j) = n.$$

### 3. Proof of Theorem 1

Let $\tilde{x}(z) = (x_o(z), x_1(z), \ldots, x_n(z)):C \to C^{n+1} - \{0\}$ be a reduced representation of $x$. We put

$$|\tilde{x}(z)| = (\sum_{j=0}^{m} |x_j(z)|^2)^{1/2},$$

$$|\tilde{x}(z)|_s = \max \{|x_j(z)|; 0 \leq n\},$$

$$T_s(r) = \frac{1}{2\pi} \int_0^{2\pi} \log|\tilde{x}(re^{i\theta})|_s \, d\theta - \log|\tilde{x}(0)|_s.$$

Then from [12; p. 105] we have

$$T(r) = \frac{1}{2\pi} \int_0^{2\pi} \log|\tilde{x}(re^{i\theta})| \, d\theta - \log|\tilde{x}(0)|. \qquad (3.1)$$

Since

$$|\tilde{x}(z)|_s \leq |\tilde{x}(z)| \leq (n + 1)|\tilde{x}(z)|_s,$$

we have

$$T(r) = T_s(r) + O(1). \qquad (3.2)$$

Hence it follows from [3; p. 10] that

$$T(r, x_i/x_j) < T(r) + K \ (i \neq j). \qquad (3.3)$$

where $K$ is a constant.

We shall now prove our Theorem 1. From $|\tilde{x}(z)|$

$\leq \sum_{j=0}^{n} |x_j(z)|$ and our assumptions

$$T(r, x_j) = o(T(r)) \ r \to \infty \ (j = 1, 2, \ldots, n), \qquad (3.4)$$

we have

$$T(r) \leq \sum_{j=0}^{n} T(r, x_j) + \log(n + 1) + \log|\tilde{x}(0)|$$

$$= T(r, x_o) + o(T(r)) \ r \to \infty$$

On the other hand it follows from (3.3) that for $j \neq 0$,

$$T(r, x_0) \leqq T(r, x_j) + T(r, x_0/x_j)$$

$$\leqq o\,(T(r)) + T(r) + K \quad r \to \infty$$

Hence we obtain

$$T(r) \sim T(r, x_0) \quad r \to \infty. \tag{3.5}$$

We set $A = \{a_\nu\}_{\nu=1}^{q}$, $a_\nu = [\bar{a}_{\nu 0}, \bar{a}_{\nu 1}, \ldots, \bar{a}_{\nu n}]$, $F_\nu(z)$
$= \sum\limits_{j=0}^{n} a_{\nu j} x_j(z)$ and $b_\nu(z) = - \sum\limits_{j=1}^{n} a_{\nu j} x_j(z)$.  Here we may assume without loss of

generality that $\bar{a}_{\nu 0} = 1$ or $\bar{a}_{\nu 0} = 0$.  It follows from (3.4) and (3.5) that for

$\nu = 1, \ldots, q$

$$T(r, b_\nu) = o(T(r)) \doteq o(T(r, x_0)) \quad r \to \infty \tag{3.6}$$

Since A is in general position, the number of elements belonging to the set

$A_0 = \{a_\nu \varepsilon A; \ \bar{a}_{\nu 0} = 0\}$ is at most n.  We may put $A_0 = \{a_\nu\}_{\nu=1}^{p}$, where $0 \leqq p \leqq n$.

Then we have $F_\nu(z) = -b_\nu(z)$ for $\nu = 1, 2, \ldots, p$.  Hence (3.6) implies for

$\nu = 1, \ldots, p$,

$$\delta\,(a_\nu) = 1 - \limsup_{r \to \infty} \frac{N(r, 0, F_\nu)}{T(r)} \tag{3.7}$$

$$= 1 - \limsup_{r \to \infty} \frac{N(r, 0, b_\nu)}{T(r)}$$

For $a_\nu \varepsilon A - A_0$, remarking $F_\nu(z) = x_0(z) - b_\nu(z)$ and (3.5), we have for $\nu = p + 1, \ldots, q$.

$$\delta(a_\nu) = 1 - \limsup_{r \to \infty} \frac{N(r, 0, F_\nu)}{T(r)} \tag{3.8}$$

$$= 1 - \limsup_{r \to \infty} \frac{N(r, b_\nu(z), x_0)}{T(r, x_0)} = \delta(b_\nu(z), x_0),$$

where $N(r, b_\nu(z), x_0) = N(r, 0, x_0(z) - b_\nu(z))$ and $\delta(b_\nu(z), x_0)$ is the

deficiency of $b_\nu(z)$ with respect to $x_0(z)$.

It follows from the definition of the index of degeneracy  that there are
$\lambda$ independent, linear relations with constant coefficients among entire
functions $x_0(z), x_1(z), \ldots, x_n(z)$.  But we deduce from growth conditions (3.4)
and (3.5) that its each linear relation contains no term of $x_0(z)$.  So we have
$\lambda$ independent, linear relations among $x_1(z), x_2(z), \ldots, x_n(z)$.  Hence for each

fixed integer $\ell (> p)$, the number of elements belonging to the set $B_\ell = \{\nu > p; b_\nu(z) \equiv b_\ell(z)\}$ is at most $\lambda + 1$, because $\ell$ itself belongs to $B_\nu$. Therefore taking (3.8) into account we have

$$\sum_{\nu \in B_\ell} \delta(a_\nu) \leqq \lambda + 1 \; \delta(b_\ell(z), x_o). \tag{3.9}$$

We here suppose that $b_{p+1}^*(z), \ldots, b_t^*(z)$ $(t \leqq q)$ are mutually distinct entire functions in $\{b_\nu(z)\}_{\nu=p+1}^q$. Then it follows from the result of Yang [13; Cor. 3] that

$$\sum_{\ell=p+1}^t \delta(b_\ell^*(z), x_o) \leqq 1 - \sin \pi\mu,$$

because $x_o(z)$ is an entire function of lower order $\mu(1/2) < \mu < 1$ and from (3.6) we have $T(r, b_\ell^*) = o(T(r, x_o))$ $r \to \infty$ for $\ell = p+1, \ldots, t$. Hence combining this with (3.9) we obtain

$$\sum_{a \in A - A_o} \delta(a) \leqq (\lambda - 1) \sum_{\ell=p+1}^t \delta(b_\ell^*(z), x_o)$$

$$\leqq (\lambda + 1)(1 - \sin \pi\mu)$$

and so from (3.7)

$$\sum_{a \in A} \delta(a) \leqq p + (\lambda + 1)(1 - \sin \pi\mu)$$

$$\leqq n + (\lambda + 1)(1 - \sin \pi\mu),$$

which is Theorem 1.

4.   Proof of Theorem 2.

Since $\lambda = n = 1$, there is a reduced representation $\tilde{x} = (x_o, x_1, \ldots, x_n)$: $C \to C^{n+1} - \{0\}$ of $x$ such that

$$x_j(z) = \beta_j x_o(z) + \gamma_j x_1(z), \quad j = 2, 3, \ldots, n, \tag{4.1}$$

where $\beta_j$ and $\gamma_j$ are constants. We put $A = \{a_\nu\}_{\nu=1}^q$, $a_\nu = [\bar{a}_{\nu 0}, \bar{a}_{\nu 1}, \ldots, \bar{a}_{\nu n}]$ and $F_\nu(z) = \sum_{j=0}^n a_{\nu j} x_j(z)$. Then it follows from (4.1) that

$$F_\nu(z) = (a_{\nu 0} + \sum_{j=2}^n a_{\nu j} \beta_j) x_o(z) + (a_{\nu 1} + \sum_{j=2}^n a_{\nu j} \gamma_j) x_1(z) \tag{4.2}$$

$$u \, x_o(z) + v \, x_1(z).$$

(3.3) implies

$$T(r, x_1/x_o) \leqq T(r) + K.$$

On the other hand we deduce from (3.1) and (4.1) that

$$T(r) \leqq \frac{1}{2\pi} \int_o^{2\pi} \log \left(|x_o(re^{i\theta})|^2 + |x_1(re^{i\theta})|^2\right)^{1/2} d\theta$$

$$+ O(1) = T(r, x_1/x_o) + O(1).$$

Hence we have

$$T(r) = T(r, x_1/x_o) + O(1) \quad r \to \infty \tag{4.3}$$

Now put $c_\nu = -u_\nu/v_\nu$ if $v_\nu \neq 0$ and $c_\nu = \infty$ if $v_\nu = 0$. Remarking that $x_o(z)$ and $x_1(z)$ have no common zero, we can deduce that

$$n(r, 0, F_\nu) = n(r, c_\nu, x_1/x_o)$$

and consequently, taking (3.7) and (4.3) into account,

$$\delta(a_\nu) = \delta(c_\nu, x_1/x_o). \tag{4.4}$$

Next, if $c_\nu = c_\ell \neq \infty$, that is, $u_\nu/v_\nu = u_\ell/v_\ell$, then from (4.2) we have $F_\nu(z) = (v_\nu/v_\ell)F_\ell(z)$. If $c_\nu = c_\ell = \infty$, then $u_\nu \neq 0$, $u_\ell \neq 0$ and $F_\nu(z) = (u_\nu/u_\ell)F_\ell(z)$. Hence, since $\lambda = n - 1$, for each fixed integer $\ell$ the number of elements of the set $D_\ell = \{\nu; c_\nu = c_\ell\}$ is at most n. Therefore, it follows from (4.4) that

$$\sum_{\nu \in D\ell} \delta(a_\nu) \leqq n \, \delta(c_\ell, x_1/x_o).$$

Let $c_1^*, c_2^*, \ldots, c_t^*$ be mutually distinct values in $\{c_\nu\}_{\nu=1}^q$. Then it follows from Theorem A that

$$\sum_{\ell=1}^t \delta(c_\ell^*, x_1/x_o) \leqq 2 - \sin \pi\mu$$

Hence we obtain

$$\sum_{a \in A} \delta(a) \leqq n \sum_{\ell=1}^t \delta(c_\ell^*, x_1/x_o) \leqq n \, (2 - \sin \pi\mu),$$

which proves Theorem 2.

5.   Proof of Theorem 3.

To construct a holomorphic curve proving Theorem 3, we use Toda's idea [11] with Lindelöf's function $L_\mu(z) = \prod_{n=1}^\infty (1 + z/n^{1/\mu})$ of lower order $\mu$ $(1/2 < \mu < 1)$. With respect to $L_\mu(z)$ it is well known that $\delta(0, L_\mu) = 1 - \sin \pi\mu$.

Let $\beta_1, \ldots, \beta_{n+\lambda+1}$ be $n + \lambda + 1$ distinct finite complex numbers.  We define

entire functions $f_0(z), f_1(z), \ldots, f_n(z)$ by

$$F(z, w) \equiv f_0(z)w^n + f_1(z)w^{n-1} + \ldots + f_{n-1}(z)w + f_n(z) \tag{5.1}$$

$$= L_\mu(z) \prod_{i=1}^{n} (w - \beta_i) + \prod_{j=0}^{\lambda} (w - \beta_{n+1+j}) \sum_{k=1}^{n-\lambda} B_k(z) \sum_{\substack{m=1 \\ m \neq k}}^{n-\lambda} (w - \beta_m),$$

where $B_k(z) = z^{k-1}/(\prod_{j=0}^{\lambda} (\beta_k - \beta_{n+1+j}) \prod_{\substack{m=1 \\ m \neq k}}^{n-\lambda} (\beta_k - \beta_m))$.  And put $F_i(z) = F(z, \beta_i)$

$(i = 1, 2, \ldots, n + \lambda + 1)$.  Then we have

$$\left\{ \begin{array}{ll} F_i(z) = z^{i-1} & (i = 1, 2, \ldots, n - \lambda), \\[1.5em] F_{n-\lambda+\ell}(z) = \displaystyle\sum_{k=1}^{n-\lambda} c_{\ell k} z^{k-1} = \sum_{k=1}^{n-\lambda} c_{\ell k} F_k(z) & \\[1em] & (\ell = 1, \ldots, \lambda), \\[1.5em] F_{n+1+p}(z) = d_p L_\mu(z) & (p = 0, 1, \ldots, \lambda), \end{array} \right. \tag{5.2}$$

where $c_{\ell k} = \displaystyle\prod_{j=0}^{\lambda} \{(\beta_{n-\lambda+\ell} - \beta_{n+1+j})/(\beta_k - \beta_{n+1+j})\} \prod_{\substack{m=1 \\ m \neq k}}^{n-\lambda}$

$\{(\beta_{n-\lambda+\ell} - \beta_m)/(\beta_k - \beta_m)\}$ and $d_p = \displaystyle\prod_{i=1}^{n} (\beta_{n+1+p} - \beta_i)$.

We now put

$$x_\nu(z) = F_{\nu+1}(z), \quad \nu = 0, 1, \ldots, n. \tag{5.3}$$

Then we deduce from (5.2) and (5.3) that the holomorphic curve $x:C \to P_nC$

induced by $\tilde{x}(z) = (x_0(z), x_1(z), \ldots, x_n(z))$ satisfies the conditions in our

Theorem 3 with $T(r) \sim T(r, L_\mu)$ $(r \to \infty)$.  Let $R$ be the regular matrix such that

$\tilde{x}(z) = \tilde{f}(z)R$, where $\tilde{f}(z) = (f_0(z), f_1(z), \ldots, f_n(z))$.  Further we put $\tilde{b}_j =$

$(\bar{\beta}_j^n, \bar{\beta}_j^{n-1}, \ldots, \bar{\beta}_j, 1)$, $\tilde{a}_j = \tilde{b}_j(R^{-1})*$, $b_j = \pi(\tilde{b}_j)$ and $a_j = \pi(\tilde{a}_j)$ $(j = 1, 2, \ldots,$

$n + 1 + \lambda)$, where $*$ is the adjoint operator of matrices and $\pi$ is the canonical

projection of $C^{n+1} - \{0\}$ into $P_nC$.  Since $\{b_j\}$ is in general position, $\{a_j\}$ is

also in general position.  And (5.1) implies

$$\langle \tilde{x}(z), \tilde{a}_j \rangle = \tilde{x}(z)\tilde{a}_j* = \tilde{f}(z)\tilde{b}_j* = \langle \tilde{f}(z), \tilde{b}_j \rangle = F_j(z) \quad (j = 1, 2, \ldots,$$

$$n + \lambda + 1).$$

Hence it follows from (5.2) that

$$\delta\,(a_j) = 1 \quad (j = 1, \ldots, n)$$

and

$$\delta(a_{n+k}) = \delta(0, L_\mu) = 1 - \sin \pi\mu \quad (k = 1, \ldots, \lambda + 1).$$

Thus Theorem 3 is proved.

   6.   Proof of Theorem 4.

   In the first place let us recall the definition of the spread $\sigma(a)$ of $a\varepsilon P_n C$. Let $\{r_m\}$ be a sequence of Polya peaks of order $\mu$ of x and $\Lambda(r)$ a positive function which increases unboundedly satisfying

$$\Lambda(r) = o\ (T(r))\quad r \to \infty. \tag{6.1}$$

We define the set of arguments $E_\Lambda\ (r,\ a) \subset (-\pi,\ \pi)$ by

$$E_\Lambda\ (r,\ a) = \{\theta; |<x(re^{i\theta}),\ a>| < e^{-\Lambda(r)}\}$$

for $a\varepsilon P_n C$ satisfying $<x(z),\ a > \not\equiv 0$ and let

$$\sigma_\Lambda(a) = \lim_{m \to \infty} \inf\ \text{means}\ E_\Lambda(r_m,\ a),$$

$$\sigma(a) = \inf_\Lambda\ \sigma_\Lambda(a),$$

where the infimum is taken over all function $\Lambda(r)$ satisfying (6.1).  Then we have the following spread relation

   LEMMA 1 ([9], [8]).  Let $x : C \to P_n C$ be a holomorphic curve of positive, finite lower order $\mu$.  Then

$$\sigma(a) \geqq \min\ \{2\pi,\ (4/\mu)\ \sin^{-1}\ (\delta(a)/2)^{1/2}\}$$

for every $a\varepsilon P_n C$ satisfying $< x(z),\ a > \not\equiv 0$.

   We can next deduce from Krutin [7,§7] that

   LEMMA 2.  Let $x : C \to P_n C$ be a holomorphic curve, A a finite system of $P_n C$ in general position and $A_{n+1}$ an arbitrary subsystem of A consisting of $n + 1$ elements.  Then we have

$$\bigcap_{a\varepsilon A_{n+1}} E_\Lambda\ (r,\ a) = \phi$$

and so

$$\sum_{a\varepsilon A} \sigma(a) \leqq 2n\pi.$$

We are now ready to prove our Theorem 4.  Put $A_1 = A - \{a_j\}_{j=1}^{n-1}$.  From our

124     K. NIINO

assumptions we have $\sigma(a_j) = 2\pi$ $(j = 1, \ldots, n - 1)$. Hence Lemma 2 implies

$\sum\limits_{a \in A_1} \sigma(a) \leqq 2\pi$. Therefore it follows from the spread relation (Lemma 1) and

the argument in Edrei [5; pp. 442-443] that

$$\sum_{a \in A_1} \sigma(a) \leqq 2 - \sin \pi\mu$$

and so

$$\sum_{a \in A} \delta(a) \leqq n + 1 - \sin \pi\mu,$$

Since $\delta(a_j) \leqq 1$ $(j = 1, \ldots, n - 1)$. This proves our Theorem 4.

   7. Proof of Theorem 5.

   We may assume from Theorem B that $\mu > 1/2$. We put $A_1 = \{a \in A; \sigma(a) = 2\pi\}$.

Then it follows from Lemma 2 and our assumption that the number of elements of

$A_1$ is at most n and at least one. Since $\lambda = n - 1$, we can here use the same

notations as the proof of Theorem 2 in 4. Further we may assume without loss

of generality that $a_0 \in A_1$, $a_1 \in A_1$, $a_0 = [1, 0, \ldots, 0]$, $a_1 = [0, 1, 0, \ldots, 0]$

and (4.1) and (4.2) are valid. For if $x_0(z)$ is proportional to $x_1(z)$ then

$a_0 \in A_1$. Hence $x_0(z)$ and $x_1(z)$ are linearly independent.

   We define $x^*(z) = \max\{|x_0(z)|, |x_1(z)|\}$ and $E'_\Lambda(r, a_\nu) = \{\theta; \log x^*$

$(re^{i\theta}) - \log|F_\nu(re^{i\theta})| > \Lambda(r)\} \subset (-\pi, \pi]$. From the reasoning of [9; p. 363]

we deduce

$$\inf_\Lambda \liminf_{m \to \infty} E'_\Lambda(r_m, a_\nu) = \sigma(a_\nu).$$

Hence we can consider $E'_\Lambda(r, a_\nu)$ instead of $E_\Lambda(r, a_\nu)$.

   From the definition we have $E'_\Lambda(r, a_1) = \{\theta; \log|x_0(re^{i\theta})| - \log|x_1$

$(re^{i\theta})| > \Lambda(r)\}$. Hence, if $\theta \in E'_\Lambda(r, a_1)$, then $|x_0(z)| > e^{\Lambda(r)}|x_1(z)|$, $z = re^{i\theta}$. Further if $u_\nu \neq 0$, we obtain for $z = re^{i\theta}$

$$\log x^*(z) - \log|F_\nu(z)| = \log|x_0(z)| - \log|u_\nu x_0(z) + v_\nu x_1(z)|$$

$$\leqq - \log(|u_\nu| - |v_\nu|e^{-\Lambda(r)}).$$

Therefore, $\theta \in E'_\Lambda(r, a_1)$ and $u_\nu \neq 0$ yield $\theta \notin E'_\Lambda(r, a_\nu)$ for sufficiently large

r. Since $\sigma(a_1) = 2\pi$, $u_\nu \neq 0$ implies that $\sigma(a_\nu) = 0$ and so $a_\nu \notin A_1$ and that

$\delta(a_\nu) = 0$ by Lemma 1. On the other hand, $u_\nu = 0$ implies $F_\nu(z) = v_\nu x_1(z)$ and

so $a_\nu \in A_1$. Thus it follows that $\delta(a) = 0$ for $a \notin A_1$. Hence we obtain

$$\sum_{a \in A} \delta(a) = \sum_{a \in A_1} \delta(a) \leqq n,$$

since the number of elements of $A_1$ is at most n.

Thus the proof of Theorem 5 is complete.

8.  Proof of Theorem 6.

Let $L_\mu(z)$ be Lindelöf's function of lower order $\mu$ $(1/2 < \mu < 1)$:

$$L_\mu(z) = \prod_{n=1}^{\infty} (1 + z/n^{1/\mu}).$$

Then it is well known that for $z = re^{i\theta}$

$$\log|L_\mu(z)| = \frac{\pi}{\sin \pi\mu}\, r^\mu \cos \mu\theta (1 + \eta(z)) \qquad (8.1)$$

$$|\theta| \leqq \pi - \delta \quad (\delta > 0)$$

and so

$$T(r, L_\mu) = m(r, L_\mu) = \frac{1}{\mu \sin \pi\mu}\, r^\mu (1 + \eta(r)), \qquad (8.2)$$

$$N(r, 0, L_\mu) = \frac{1}{\mu}\, r^\mu (1 + \eta(r)),$$

where $\eta(z) \to 0$ $(z \to \infty,\ |\theta| \leqq \pi - \delta)$ and $\eta(r) \to 0$ $(r \to \infty)$.

For any given $\varepsilon > 0$ we choose an integer p satisfying

$$p > \max \{n, ((n - \lambda)(n - \lambda - 1)/2 + \lambda)/\varepsilon\}. \qquad (8.3)$$

We put $\tilde{x}(z) = (x_o(z), x_1(z), \ldots, x_n(z))$, $x_o(z) = 1$, $x_j(z) = L_\mu(z)^j$

$(j = 1, \ldots, n - \lambda - 1)$, $x_j(z) = x_1(z) = L_\mu(z)$ $(j = n - \lambda, \ldots, n - 1)$ and

$x_n(z) = L_\mu(z)^p$.  Then it easily follows that

$$\log|\tilde{x}(z)|_s = \log\overset{+}{|x_n(z)|} = p\, \log\overset{+}{|L_\mu(z)|}$$

and so

$$T(r) = \frac{p}{\mu \sin \pi\mu}\, r^\mu (1 + \eta(r)). \qquad (8.4)$$

We also have for $z = re^{i\theta}$ and $j = 1, \ldots, n - \lambda - 1$

$$\log|\tilde{x}(z)|_s - \log|x_j(z)|$$

$$= \begin{cases} ((p - j)/\sin \pi\mu)\, r^\mu \cos \mu\theta(1 + \eta(z)), & \theta \in G_1, \\ (-j/\sin \pi\mu)\, r^\mu \cos \mu\theta(1 + \eta(z)), & \theta \in G_2, \end{cases}$$

where $G_1 = \{\theta;\ |\theta| \leqq \pi - \delta,\ \cos \mu\theta > 0\}$, $G_2 = \{\theta;\ |\theta| \leqq \pi - \delta,\ \cos \mu\theta < 0\}$.

Hence letting $a_j = [0, \ldots, 0, 1, 0, \ldots, 0]$ ($j$ - th spot), $j = 0, 1, \ldots, n$

and taking $x_j(z) = x_1(z)$ ($j = n - \lambda, \ldots, n - 1$) into account, we deduce from

the reasoning in [9; p. 363] that $\sigma(a_j) = 2\pi$ ($j = 1, \ldots, n - 1$), $\sigma(a_o) = \pi/\mu$

and $\sigma(a_n) = 2\pi - (\pi/\mu)$.  Further it follows from (8.2), (8.3) and (8.4) that

$$\sum_{j=0}^{n} \delta(a_j) = 1 + \sum_{j=1}^{n-\lambda-1} (1 - (j/p) \sin \pi\mu) + \lambda(1 - (1/p) \sin \pi\mu) + 1 - \sin \pi\mu >$$

$$n + 1 - (1 + \varepsilon) \sin \pi\mu.$$

Thus the proof of Theorem 6 is complete.

    9.  Proof of Theorem 7.

    In order to prove Theorem 7 we need a lemma.  From the reasoning of
Clunie [4], especially the proof of his lemma 8 we can deduce

    LEMMA 3.  Let $\phi(r)$ be increasing and convex in $\log r$ with $\phi(r) \neq 0$
($\log r$) ($r \to \infty$).  Then there is an entire function $f(z)$ such that $T(r, f) \sim \phi$
$(r)$ ($r \to \infty$) and $\sigma(\infty, f) = 2\pi$, where $\sigma(\infty, f)$ is the spread of $\infty$ for the entire
function $f(z)$.

    We shall now prove Theorem 7.  Let $f(z)$ be the entire function obtained
by Lemma 3 for the given $\phi(r)$.  We put $x_o(z) = f(z)$, $x_1(z) = \ldots = x_n(z) = 1$.
Let $x : C \to P_n C$ be the holomorphic curve induced by $\tilde{x}(z) = (x_o(z), x_1(z), \ldots, x_n(z))$.  Then we have $\log|\tilde{x}(z)|_s = \log^+|f(z)|$ and from (3.2)

$$T(r) \sim T(r, f) \sim \phi(r) \quad (r \to \infty).$$

    We next put $a_j = [0, \ldots, 0, 1, 0, \ldots, 0]$ ($j$ - th spot).  Then we have
for $j = 1, 2, \ldots, n$

$$E'_\Lambda(r, a_j) = \{\theta;\ \log|\tilde{x}(re^{i\theta})|_s - \log|x_j(re^{i\theta})| > \Lambda(r)\}$$

$$= \{\theta;\ \log^+|f(re^{i\theta})| > \Lambda(r)\} = E_\Lambda(r, \infty; f).$$

Hence we have $\sigma(a_j) = \sigma(\infty, f) = 2\pi$ ($j = 1, 2, \ldots, n$).  And it clearly follows
that $\delta(a_j) = 1$ ($j = 1, 2, \ldots, n$).

    Thus the proof of Theorem 7 is complete.

## REFERENCES

[1] Ahlfors, L.V., The theory of meromorphic curves. Acta Soc. Sci. Fenn. Ser. A, 3 (1941), no. 4, 31 pp.

[2] Baernstein II, A., Proof of Edrei's spread conjecture. Proc. London Math. Soc. (3) 26 (1973), 418-434.

[3] Cartan, H., Sur les zeros des combinaisons lineaired de p functions holomorphic donnees. Mathematica, Cluj. 7 (1933), 5-29.

[4] Clunie, J., On integral functions having prescribed asymptotic growth. Canadian J. Math. 17 (1965), 396-404.

[5] Edrei, A., Solution of the deficiency problem for functions of small lower order. Proc. London Math. Soc. (3) 26 (1973), 435-445.

[6] Hayman, W.K., Meromorphic functions. Clarendon Press, Oxford 1964.

[7] Krutin, V.I., On the magnitudes of the positive deviations and of the defects of entire curves of finite lower order. Math. USSR Izvestija 13 (1979), 307-334.

[8] Krytov, A.V., Deficiencies of entire functions of finite lower order. Ukrainian Math. J. 31 (1979), 210-214.

[9] Niino, K., Spread relation and value distribution in an angular domain of holomorphic curves. Kōdai Math. Sem. Rep. 28 (1977), 361-371.

[10] Petrenko, V.P. and M. Hussain, The defects of entire curves of lower order $\lambda < 1$ (Russian). Tear. Funkcii Funkcional. Anal. i Prilozen, 24 (1975), 128-138.

[11] Toda, N., Sur quelques combinaissons lineaires exceptionnelles au sens de Nevanlinna, V. Nagoya Math. J. 66 (1977), 37-52.

[12] Wu, H., The equidistribution theory of holomorphic curves. Ann. of Math. Studies, No. 64, Princeton Univ. Press, Princeton, 1970.

[13] Yang, L., Deficient functions of meromorphic functions Sci. Sinica 24 (1981), 1179-1189.

FACULTY OF TECHNOLOGY
KANAZAWA UNIVERSITY
2-40-20, KODATSUNO,
KANAZAWA 920, JAPAN

Contemporary Mathematics
Volume **25**, 1983

A FULLY GENERAL NEVANLINNA N-SMALL FUNCTION THEOREM

AND A SOMETIMES EFFECTIVE THUE-SIEGEL-ROTH-SCHMIDT THEOREM FOR SOLUTIONS

TO LINEAR DIFFERENTIAL EQUATION

by

Charles F. Osgood

R. Nevanlinna in addition to proving his famous result about the distribution of (constant) values of meromorphic functions also proved the Three Small Function Theorem.  The Three Small Function Theorem can be viewed as bounding the approximation (on different subarcs of circles centered at the origin) of up to three small meromorphic functions by one large meromorphic function.  Chuang extended this result [1], obtaining, for each natural number n, an n-small function theorem that retains the original Nevanlinna bound of $(2+o(1))\ T(r,f)$ when f is entire.  However, if f has more than $o(T(r,f))$ poles his bound is, generally, larger than the Nevanlinna bound.

In a paper which will soon be submitted, I prove an n-Small Function Theorem which retains the Nevanlinna bound for all meromorphic functions f. The restriction that f must have fewer than the expected number of poles is therefore removed.  The investigations which led to the proof were directed originally to proving an effective Thue-Siegel-Roth-Schmidt Theorem bounding the approximation of algebraic functions (of a complex variable z) by rational functions of z.  (See [2,3,4].)  The new proof uses auxiliary differential equations especially constructed for this purpose.  The proof extends in a natural manner to a (sometimes effective) Thue-Siegel-Roth-Schmidt Theorem for solutions of linear differential equations with polynomial coefficients.  (See [5].)  Finally, utilizing Nevanlinna's Lemma on the Logrithmic Derivative, the proof extends to Nevanlinna's n-Small Function Theorem (as well as to results involving more than one large function).

In the work of Nevanlinna and Chuang linear differential equations were used, at least implicitly. The auxiliary differential equation constructed in my proof is nonlinear. It is constructed by what amount to the following procedure (although the actual formal proof is less intuitive): In the proof of the original Thue-Siegel-Roth-Schmidt Theorem (for algebraic numbers), a large set of good approximations were assumed to exist and a crucially auxiliary polynomial P was constructed, in many unknowns $X_1,\ldots,X_N$ which among its other important properties, was nonzero when evaluated at a vector of these distinct good approximations. If a function g(z) is well approximated by f(z) then (for a set of N different independent variables $z_1$, $z_2$,...,$z_N$) each $g(z_i)$ is well approximated by $f(z_i)$. Guided by the situation with algebraic numbers, it is possible to construct here an analogue of the crucially important polynomial P there, which does not vanish when evaluated at the vector $(f(z_i))$. At $z=z_1=\ldots\ldots=z_N$ we can extract a partial derivative which does not vanish. Our auxiliary polynomial P becomes an auxiliary nonlinear differential polynomial P which does not vanish when evaluated at $f(z)$. This is the crucial new idea in my proof.

## References

1. Chuang, Chi-Tai( Zhuang, Qitai), On the Distribution of Values of Meromorphic Functions, Kexue Tongbao Vol. 25, No. 8, Aug. 1980.

2. C.F. Osgood, An effective lower bound on the "Diophantine approximation" of algebraic functions by rational functions, Mathematika 20 (1973),4-15.

3. ――――――, Effective bounds on the "Diophantine approximation" of algebraic functions over fields of arbitrary charcteristic and applications to differential equations, Proc. Koninkl Nederl. Akademie Van Wetens. Ser. A 78(1975), 105-119(errata 78, No. 5).

4. ――――――, An effective lower bounds on the "diophantine" approximation of algebraic functions by rational functions(II), CONTRIBUTIONS TO ALGEBRA: A collection of papers Dedicated to Ellis Kolchin, Academic Press,321-327.

5. ――――――, Concerning a Possible "Thue-Siegel-Roth Theorem" for Algebraic Differential Equations, from Number Theory and Algebra, Academic Press,223-234.

Naval Research Laboratory
Washington, D.C. 20375

Contemporary Mathematics
Volume 25, 1983

## MEROMORPHIC FUNCTIONS SATISFYING
## A DIFFERENTIAL EQUATION

by

Ranjan Roy and S.M. Shah

### INTRODUCTION

We consider meromorphic solutions of the differential equation (DE)

$$L_n(W,z,P) \equiv L_n(W,P) = P_0(z)W^{(n)} + P_1(z)W^{(n-1)} + \ldots + P_n(z)\, W = 0 \tag{1}$$

with polynomial coefficients $P_j$ and $P_0 \neq 0$. Let $W(z)$ be meromorphic in the complex plane $|z| < \infty$ and satisfy the DE (1). We obtain bounds for Nevanlinna characteristic $T(r,W)$ and extend some known results for entire solutions of (1) ([1], [5], [8]) in Section 1. In Section 2 we seek for vector-valued solutions of the DE

$$L_n(W,z,Q) \equiv L_n(W,Q) = W^{(n)} + Q_1(z)W^{(n-1)} + \ldots + Q_n(z)W = 0 \tag{2}$$

where $Q_i$ are mxm matrices with rational entries.

### SECTION 1

Let $W$ be meromorphic and satisfy the DE (1) where $P_j$ are all polynomials. It is well-known that the singularities of the solutions $W$ are at the singularities of $P_j(z)/P_0(z)$ $(j = 1,2,\ldots,n)$ and these are finite in number. Hence $W(z) = E(z)/Q(z)$ where $E$ is entire and $Q$ a polynomial whose zeros form a subset of the zeros of $P_0$. With this notation we have

LEMMA 1. The function $E(z)$ satisfies a DE of the form

$$Q_0(z)E^{(n)}(z) + Q_1(z)E^{(n-1)}(z) + \ldots + Q_n(z)E(z) = 0 \tag{3}$$

where $Q_i$ are polynomials. Moreover if $\deg P_0 \geq \max_{1 \leq j \leq n}\{\deg P_j\}$ then $\deg Q_0 \geq \max_{1 \leq j \leq n}\{\deg Q_i\}$. If $\deg P_0 = \deg P_k$ for some $k$ $(1 \leq k \leq n)$ then $\deg Q_0 = \deg Q_k$.

PROOF. We prove by induction that if $W = \dfrac{E}{Q}$ then

$$W^{(n)} = \frac{Q^n E^{(n)} + R_1 E^{(n-1)} + R_2 E^{(n-2)} + \ldots + R_n E}{Q^{n+1}}$$

where $R_1,\ldots,R_n$ are polynomials whose degrees are strictly less than the degree of $Q^n$. For $n=1$, the result is true, since

$$W' = \frac{QE' - Q'E}{Q^2} .$$

Suppose that

$$W^{(n-1)} = \frac{Q^{n-1}E^{(n-1)} + R_1 E^{(n-2)} + \ldots + R_{n-1}E}{Q^n}.$$

Differentiating the equation, we get

$$W^{(n)} = \frac{Q^n(Q^{n-1}E^{(n)} + (n-1)Q^{n-2}E^{(n-1)}Q' + \ldots) - nQ^{n-1}Q'(Q^{n-1}E^{(n-1)} + \ldots)}{Q^{2n}}$$

$$= \frac{Q^n E^{(n)} + R_1 E^{(n-1)} + \ldots + R_n E}{Q^{n+1}}$$

which is of the required form.  Substituting the values of $W^{(i)}$ in equation (1), we have

$$\frac{P_0(Q^n E^{(n)} + R_1 E^{(n-1)} + \ldots)}{Q^{n+1}} + \frac{P_1(Q^{n-1}E^{(n-1)} + R_1' E^{(n-2)} + \ldots)}{Q^n} + \ldots \frac{P_n E}{Q} = 0$$

so that

$$P_0 Q^n E^{(n)} + P_0 R_1 E^{(n-1)} + \ldots + P_1 Q^n E^{(n-1)} + \ldots + P_n Q^n E = 0.$$

The statements of the lemma follow from this.  Suppose now that $P_0 = \deg P_0 \geq$ $\max\limits_{1 \leq i \leq n} \{\deg P_i\}$.  Let $P_i(z) = a_i z^{P_0}$ + terms of degree less than $P_0$ $(i=0,1,\ldots,n)$, $a_0 \neq 0$.  It is possible that $a_i = 0$ for $i=1,2,\ldots,n$.  Suppose that $Q(z) = bz^q + \cdots$.  It is clear from the proof of Lemma 1 that the polynomials $Q_i(z)$ in (3) are of the form

$$Q_i(z) = b^n a_i z^{P_0 + nq} + \cdots.$$

With this notation, we can state

THEOREM 1.  If $W$ is a meromorphic solution of the DE (1), and $\deg P_0 \geq \max\limits_{1 \leq i \leq n} \deg\{P_i\}$, then

$$\limsup_{r \to \infty} \frac{T(r,W)}{r} \leq T \leq \max\left\{ c, \frac{1}{|a_0|}\left( \sum_{k=1}^{n} \frac{|a_k|}{c^{k-1}} \right) \right\} \tag{4}$$

where $c$ is any positive number and $T = \limsup\limits_{r \to \infty} \dfrac{\log M(r,E)}{r}$.  If $\deg P_0 > \max\limits_{1 \leq i \leq n} (\deg P_i)$ then

$$\limsup_{r \to \infty} \frac{T(r,W)}{r} = 0.$$

The proof depends on the following lemma [8].

LEMMA 2.  Let $E(z)$ be the function of Lemma 1 and suppose that $\deg Q_0 \geq \max\limits_{1 \leq i \leq n}$ $(\deg Q_i)$.  Then $E(z)$ satisfies

$$\limsup_{r \to \infty} \frac{\log M(r,E)}{r} = T \leq \max \left\{ c, \left|\frac{1}{a_0}\right| \left( \sum_{k=1}^{n} \frac{|a_k|}{c^{k-1}} \right) \right\}. \tag{5}$$

where $c$ is any positive number.

PROOF OF THEOREM 1.  We have

$$T(r,W) \leq T(r,E) + 0(\log r)$$

$$\leq \log M(r,E) + 0(\log r). \tag{6}$$

The conclusion follows from Lemma 2.  Note that if $\deg P_0 > \max\limits_{1 \leq i \leq n} (\deg P_i)$ then $a_k = 0$ for $k = 1, 2, \ldots, n$.

COROLLARY 1.1.  If $W$ has maximum deficiency sum

$$\sum_a \delta(a,W) = 2$$

and satisfies the DE (1), where $\deg P_0 \geq \max\limits_{1 \leq i \leq n} \{\deg P_i\}$ then

$$\limsup_{r \to \infty} \frac{T(r,W)}{r} \leq \frac{T}{\pi} \leq \frac{1}{\pi} \max \left\{ c, \frac{1}{|a_0|} \sum_{k=1}^{n} \frac{|a_k|}{c^{k-1}} \right\}. \tag{7}$$

The inequality is sharp.

To prove (7) we use (5) and a known result [9].

EXAMPLE.  $W = \exp(dz)$, $d > 0$.  Then $W$ satisfies $W' - dW = 0$, $\sum \delta(a,W) = 2$ and the extreme right term in (7) is $\frac{d}{\pi}$.  Also $\limsup\limits_{r \to \infty} \frac{T(r,W)}{r} = \frac{d}{\pi}$.

THEOREM 2.  Suppose that all solutions of (1) are meromorphic functions.

    (i)  If $\deg P_0 \geq \max\limits_{1 \leq i \leq n} \deg P_i$ then the order $\rho(W) \leq 1$ for each solution $W$.

    (ii) If $\deg P_0 < \max\limits_{1 \leq i \leq n} \deg P_i$ then there is a solution $W$ for which $\rho(W) > 1$.

PROOF.  The first part (i) is proved in Theorem 1.

    (ii) Suppose now $\deg P_0 < \max\limits_{1 \leq i \leq n} \deg P_i$.  Let $W_1, W_2, \ldots, W_n$ be the $n$ Linearly independent solutions of the DE (1).  By our hypothesis these functions are all meromorphic.  Let $Q(z)$ be a polynomial which has zeros of correct multiplicity at the poles of the solutions $W_i$, that is $W_i = E_i/Q$ $(i = 1, 2, \ldots, n)$ and $E_i$ are all entire.  Clearly $E_i$ $(i = 1, 2, \ldots, n)$ are linearly independent entire solutions of the DE (3).  Hence by Theorem 4.9 of [1] it follows that the DE (3) will have a solution E such that the order $\rho(E) > 1$.  Since $W = E/Q$, $\rho(W) > 1$.

We shall now extend the concept of bounded index (BI) to meromorphic functions with a finite number of poles.  As we observed these are the only meromorphic functions which can be solutions of (1).

DEFINITION.  Suppose $f(z)$ is a meromorphic function with a finite number of poles, so that $f(z) = E(z)/Q(z)$ where Q is a polynomial and E is entire.  We say that f is of BI if E is of BI.  Suppose that Q is a polynomial of least degree such that E is entire.  We shall say that f has index N if E has index N.

We state the following consequence of the above definition and the theorem of Hayman [3] (see also [6]).

THEOREM 3.  If $f(z) = E(z)/Q(z)$ is a meromorphic function of index N, then

$$|f(z)| \le \frac{C}{|Q(z)|} \exp\{(N+L)|z|\} \quad \text{for } |z| > R$$

where R is chosen large enough.

We may also state the following consequence of Theorems 2 and 3.

COROLLARY 2.1.  With the assumptions of Theorem 2, we have $\deg P_o \ge \max_{1 \le i \le n} \deg P_i$ if and only if all solutions are of BI.

PROOF.  If $\deg P_o \ge \max_{1 \le i \le n} \deg P_i$, then all solutions are of BI ([7][2]).  Conversely if all solutions are of BI then by Theorem 3 they are of order not exceeding one and consequently from Theorem 2 $\deg P_o \ge \max_{1 \le i \le n} \deg P_i$.

THEOREM 4.  Let $G(z)$ be a meromorphic function with a finite number of poles and of BI.  Suppose W is a meromorphic function and satisfies the DE

$$L_n(W,P) = G(z) , \qquad\qquad (8)$$

where $L_n(W,P)$ is defined in (1).  If $\deg P_o \ge \max_{1 \le i \le n} \{\deg P_i\}$ then W is of BI.

PROOF.  Since $G = g/p$, where g is an entire function and p a polynomial, equation (8) is equivalent to

$$P_o^* W^{(n)} + P_1^* W^{(n-1)} + \ldots + P_n^* W = g ,$$

and $P_k^*$ are polynomials.  Note that W has only a finite number of poles.  Write $W = E/Q$ where E is entire and Q a polynomial.  As in Lemma L, we get

$$P_o^* Q^n E^{(n)} + (P_o^* R_1 + P_1^* Q^n) E^{(n-1)} + \ldots + P_n^* E Q^n = g Q^{n+1} .$$

Now the expression on the right is an entire function of BI since g is of BI.

We use the argument in [2] to show that E, and hence W, is of BI.

## SECTION 2

We suppose that all $Q_i(z)$ in the DE (2) are m×m matrices with entries which are rational functions of $z$, bounded at $\infty$.  Let $W(z) = (W_1(z), \ldots, W_m(z))$ be a vector-valued solution of the DE (2).

Assume that $W_k(z)$ are meromorphic functions with a finite number of poles. Let $Q(z)$ be a polynomial such that $W_k(z) = E_k(z)/Q(z)$ $(k=1,2,\ldots,m)$ and $E_k(z)$ are entire functions.  As in the proof of Lemma 1, we have

$$W_k^{(n)} = \frac{E_k^{(n)} + R_1^* E_k^{(n-1)} + \ldots + R_k^* E_k}{Q}$$

where $R_k^* = P_k/Q^n$ are rational functions that vanish at $\infty$.  Let $E(z) = (E_1(z), E_2(z), \ldots, E_m(z))$.  Clearly

$$W^{(n)}(z) = \frac{E^{(n)}(z) + R_1^* E^{(n-1)}(z) + \ldots + R_n^* E(z)}{Q(z)} .$$

Substituting this and similar relations in the DE (2) we see that $E(z)$ satisfies

$$E^{(n)} + Q_1' E^{(n-1)} + \ldots + Q_n' E = 0$$

where $Q_k'$ are matrices with rational entries which are bounded at $\infty$.  Moreover, since the rational functions $R_i^*(z)$ vanish at infinity, we see that

$$\lim_{z \to \infty} Q_k(z) = \lim_{z \to \infty} Q_k'(z) = (a_{ij,k}) \quad \text{(say)} .$$

Let

$$A_k = \sup\{|a_{ij,k}|, \ 1 \le i,j \le m\}. \tag{9}$$

To state the next theorem we introduce the characteristic function of a vector-valued meromorphic function defined as follows:

$$T(r,W) = \max_{1 \le i \le m} T(r,W_i) .$$

THEOREM 5.  Let $W(z)$ be a vector-valued meromorphic function with a finite number of poles satisfying

$$L_n(W,Q) = 0 .$$

Then

$$\limsup_{r \to \infty} \frac{T(r,W)}{r} \le \max\left\{ 1, \ m \sum_{k=1}^{n} A_k \right\} .$$

where $A_k$ are as in (9).

PROOF. Let $W = E/Q$, where $Q$ is a polynomial and $E$ is a vector-valued entire function. We have shown that $E$ satisfies

$$E^{(n)} + Q_1' E^{(n-1)} + \ldots + Q_n' E = 0.$$

It is easily deduced that, if $\| E^{(j)}(z) \| = \max_{1 \leq i \leq m} \{ |E_i^{(j)}(z)| \}$, then

$$\| E^{(n)} \| \leq m \left\{ A_1 + \frac{\varepsilon}{mn} \right\} \| E^{(n-1)} \| + \ldots + m \left\{ A_n + \frac{\varepsilon}{mn} \right\} \| E \|$$

for $|z| > R > R_0(\varepsilon)$. Hence

$$\| E^{(n)} \| \leq m \left\{ \sum_{i=1}^{n} A_i + \varepsilon \right\} \max_{0 \leq i \leq n-1} \| E^{(i)} \| .$$

It can be deduced from this inequality that [5]

$$\| E(z) \| \leq A \left\{ \exp \; \max\left( 1, \; m \sum_{i=1}^{n} A_i + \varepsilon \right) |z| \right\} \; \text{for} \; |z| \geq R.$$

Therefore we may immediately deduce that

$$\limsup_{r \to \infty} \frac{\log M(r,E)}{r} \leq \max \left\{ 1, \; m \sum_{i=1}^{n} A_i \right\} ,$$

where

$$M(r,E) = \max_{|z|=r} \| E(z) \| = \max_{1 \leq i \leq m} M(r,E_i), \; \text{with} \; M(r,E_i) = \max_{|z|=r} |E_i(z)| .$$

Now

$$T(r,W_i) = T\left( r, \frac{E_i}{Q} \right) \leq \log M(r,E_i) + 0(\log r).$$

Therefore

$$T(r,W) = \max_{1 \leq i \leq m} T(r,W_i) \leq \max_{1 \leq i \leq m} (\log M(r,E_i)) + 0(\log r),$$

and consequently

$$\limsup_{r \to \infty} \frac{T(r,W)}{r} \leq \max \left\{ 1, \; m \sum_{i=1}^{n} A_i \right\} .$$

The idea of a vector-valued entire function of BI was developed and used in [4] and [5]. Thus $E(z) = (E_1(z), \ldots, E_m(z))$ is of BI if there exists an integer $N$, such that

$$\max_{0 \leq i \leq N} \frac{\| E^{(i)}(z) \|}{i!} \geq \frac{\| E^{(j)}(z) \|}{j!}$$

for all $z \; \varepsilon \; C$ and $j = 0, 1, 2 \cdots$. The least such integer $N$ is called the index of $E$. We now extend this definition to vector-valued meromorphic functions $F(z) = (f_1(z), \ldots, f_m(z))$ with a finite number of poles. Let $f_i = E_i/Q$

$(i=1,2,\ldots,m)$. We say that $F$ is of BI if $E=(E_1,\ldots,E_m)$ is of BI. If $Q$ is the polynomial of smallest degree such that all $E_i$ are entire, then we say that $F$ is of index $N$ provided $E$ is of index $N$.

We are now in a position to state

THEOREM 6. If $F$ is a vector-valued meromorphic function, with a finite number of poles, that satisfies

$$L_n(W,z,Q) = g(z)$$

where $g$ is a vector-valued entire function of BI then $F(z)$ is of BI.

PROOF. Let $F(z)=E(z)/Q(z)$, where $E$ is entire. We have seen that $E$ satisfies the DE

$$E^{(n)}+Q_1'(z)E^{(n-1)}+\ldots+Q_n'(z)E = Q(z)g(z).$$

Now $Q$ is a polynomial and so $Qg$ is an entire function of BI. Hence $E$ satisfies a DE with matrix coefficients whose entries are rational functions bounded at infinity. By Theorem 4 of [5] $E$ is a function of BI. Our definition implies that $F=E/Q$ is also of BI.

A result similar to Theorem 3 of Section 1 exists and the proof is almost identical. We state it as

THEOREM 7. If $F(z)$ is a vector-valued meromorphic function of index $N$, then

$$\| F(z)\| \le \frac{A}{|Q(z)|} \exp\{(N+1)|z|\}, \quad |z| > R.$$

Here $Q(z)$ is the polynomial required to define the index $N$.

We can use Theorem 7 to prove the following growth result.

THEOREM 8. Let $W(z)$ be a vector-valued meromorphic function, with a finite number of poles, satisfying the DE

$$L_n(W,z,Q) = 0.$$

If $q$ is the least integer such that

$$m\{(n+q-1)!A_1 + \ldots + q!A_n\} < (n+q)!$$

where $A_i$ are defined by (9), then

$$\| W(z)\| \le \frac{A}{|Q(z)|}\exp\{(n+q)|z|\} \quad \text{for } |z| > R.$$

Here $R$ has to be chosen sufficiently large and $A$ is a constant.

PROOF. Let $W(z)=E(z)/Q(z)$ where $E$ is a vector-valued entire function and $Q$ a polynomial. We have seen that $E$ satisfies the equation

$$E^{(n)}+Q_1'E^{(n-1)}+\ldots+Q_n'E = 0,$$

and moreover the numbers $A_k$ defined by (9) are the same for this equation as for $L_n(W,z,Q)=0$. We differentiate the equation for $E$ $q$ times and observe that for $|z|\geq R_o(\varepsilon)$

$$\frac{E^{(n+q)}}{(n+q)!} \leq \frac{m(A_1+\varepsilon)}{n+q} \max_{0\leq j\leq n+q-1} \frac{E^{(j)}}{j!} + \cdots$$

$$+ \frac{m(A_n+\varepsilon)}{(n+q)\cdots(q+1)} \max_{0\leq j\leq n+q-1} \frac{E^{(j)}}{j!}$$

$$+ \varepsilon \max_{0\leq j\leq n+q-1} \frac{E^{(j)}}{j!} .$$

Since $\varepsilon$ can be chosen arbitrarily small, we have for $|z|\geq R$

$$\frac{E^{(n+q)}}{(n+q)!} \leq \max_{0\leq j\leq n+q-1} \frac{E^{(j)}}{j!} .$$

It can be shown that this implies that the index of $E \leq n+q-1$. An application of Theorem 7 now gives the required result.

REMARK.  In general, if the meromorphic function $F=(f_1,\ldots,f_m)$ is of BI, it does not follow that $f_i$ are also of BI.  However, if $F$ satisfies a DE of the type discussed here then it can be shown that the functions $f_i$ are also of BI. See [5].

## REFERENCES

[1]  G.H. Fricke, R. Roy and S.M. Shah.  Bounded index, entire solutions of ordinary differential equations and summability methods.  Internat. J. Math. Sci. 4 (1981), 417-434.

[2]  G.H. Fricke and S.M. Shah.  Entire functions satisfying a linear differential equation.  Indag. Math. 37 (1975), 39-41.

[3]  W.K. Hayman.  Differential inequalities and local valences.  Pacific J. Math. 44 (1973), 117-137.

[4]  L.F. Heath.  Vector-valued entire functions of bounded index satisfying a differential equation.  J. of Res. Nat. Bur. of Standards, 83 (1978), 75-79.

[5]  R. Roy and S.M. Shah.  Vector-valued entire functions satisfying a differential equation, to appear.

[6]  S.M. Shah.  Entire functions of bounded index.  Proc. Amer. Math. Soc. 19 (1968), 1017-1022.

[7]  S.M. Shah.  Entire functions of bounded index.  Lecture Notes in Math. Springer-Verlag, Vol. 599 (1977), 117-145.

[8]  S.M. Shah.  Entire solutions of linear differential equations and bounds for growth and index numbers.  Proc. Royal Soc. Edin. Math. Section A, 94 (1983), 49-60.

[9]  S.M. Shah and H. Silverman.  Entire functions with maximum deficiency sum and their means.  Math Annalen 220 (1976), 185-192.

DEPARTMENT OF MATHEMATICS
BELOIT COLLEGE
BELOIT, WISCONSIN 53511

DEPARTMENT OF MATHEMATICS
UNIVERSITY OF KENTUCKY
LEXINGTON, KENTUCKY 40506

Contemporary Mathematics
Volume 25, 1983

# A CONVOLUTION INEQUALITY WITH APPLICATIONS IN FUNCTION THEORY

by

Matts R. Éssen, John F. Rossi, and Daniel F. Shea

We describe here a new inequality, (2) below, which relates several of the indices of Nevanlinna's value distribution theory, and then discuss some of its consequences.

Let $u(z)$ be $\delta$ - subharmonic in the plane, say $u = u_1 - u_2$ where $u_1$, $u_2$ are subharmonic. We assume $u_j(0) = 0$ for $j = 1, 2$. For fixed $r > 0$, let $\tilde{u}(re^{i\theta})$ denote the symmetric decreasing rearrangement of $u(re^{i\theta})$ on $-\pi \le \theta \le \pi$, cf. [3].

Our inequality involves Baernstein's well known *-function,

$$T^*(re^{i\theta}) = \frac{1}{2\pi} \int_{-\theta}^{\theta} \tilde{u}(re^{i\phi}) d\phi + N(r, u_2) \qquad (0 \le \theta \le \pi),$$

$$N(r, u_2) = \frac{1}{2\pi} \int_{-\pi}^{\pi} u_2(re^{i\phi}) \, d\phi,$$

and depends on the fact [3] that $T^*$ is subharmonic in the upper half plane.

Put

$$(1) \qquad T(r) = T(r,u) = \sup_{0 \le \theta \le \pi} T^*(re^{i\theta}) ,$$

$$A(r,u) = \inf_\theta u(re^{i\theta}) .$$

These growth indices are related by our basic inequality,

$$(2) \qquad T^*(r^\gamma e^{i\psi}) \le \int_0^R A(t^\gamma,u) \, K_1 \left(\frac{r}{t}, \theta\right) \frac{dt}{t} +$$

$$\int_0^R T^*(t^\gamma e^{ia}) K_2 \left(\frac{r}{t}, \theta\right) \frac{dt}{t} + c_o T(2R^\gamma) \left(\frac{r}{R}\right)^{\frac{1}{2}} ,$$

where $c_o$ is an absolute constant and

$$0 \le a < \pi, \qquad \gamma = 1 - a/\pi ,$$

$$\psi = a + \gamma\theta \ (0 < \theta \le \pi), \qquad 0 < r < R < \infty ,$$

$$K_1(r,\theta) = \frac{\gamma}{2\pi^2} \log\left(\frac{1+r+2r^{1/2} \sin \frac{1}{2}\theta}{1+r-2r^{1/2} \sin \frac{1}{2}\theta}\right) ,$$

$$K_2(r,\theta) = \frac{\sin \frac{1}{2}\theta}{2\pi} \left\{ \left(r^{1/2} + r^{-1/2} - 2\cos\frac{1}{2}\theta\right)^{-1} + \left(r^{1/2} + r^{-1/2} + 2\cos\frac{1}{2}\theta\right)^{-1} \right\} .$$

142            M.ESSÉN,J.ROSSI,AND D.SHEA

Applications are possible because of the simplicity of the kernels $K_j$, in particular because of their positivity.

Here we mention some consequences of (2) for the problem of finding sharp lower bounds for

$$(3) \qquad k(u) = \limsup_{r \to \infty} \frac{A(r,u)}{T(r,u)}$$

in terms of

$$(4) \qquad \nu = \limsup_{r \to \infty} \frac{N(r,u_2)}{T(r,u)}$$

and, say,

$$(5) \qquad \lambda = \liminf_{r \to \infty} \frac{\log T(r,u)}{\log r}$$

When $u = \log|f|$ and $f$ is meromorphic, $1 - \nu = \delta(\infty,f)$ is the usual Nevanlinna deficiency.

Here are some known sharp bounds for $k(u)$:

$$(6a) \qquad k(u) \geq -\pi\lambda \qquad\qquad (\tfrac{1}{2} \leq \lambda < \infty) \ ,$$

$$\qquad\qquad -\pi\lambda/\sin\pi\lambda \quad (0 \leq \lambda < \tfrac{1}{2})$$

$$(6b) \qquad k(u) \geq \frac{\pi\lambda}{\sin\pi\lambda} (\cos\pi\lambda-\nu) \qquad (0 \leq \lambda < \tfrac{1}{2}, \ \nu < \cos\pi\lambda)$$

$$(6c) \qquad k(u) \geq -\pi\lambda \sin(\pi-\beta)\lambda \qquad (\tfrac{1}{2} \leq \lambda \leq 1, \ \nu < \sin\pi\lambda)$$

where $\beta$ is defined by $\cos \beta\lambda = \nu$, $0 < \beta \leq \pi$.

Here (6a) is due to Govorov and Petrenko [14]. This elegant result is sharp only if $\nu$ is disregarded and $u_1$ is "small" compared to $u_2$. The inequalities (6b) of Gol'dberg and Ostrovskii [13], and (6c) of Edrei-Fuchs [10] and Essén-Shea [8], improve (6a) in case $\nu$ is known and $\lambda < 1$. Examples show that (6a) is sharp in the case $\tfrac{1}{2} \leq \lambda \leq 1$ and $\sin \pi\lambda \leq \nu \leq 1$.

The class of $\delta$-subharmonic $u(z)$ having $\lambda < 1$ is by now well studied, but the known results just mentioned give no sharp estimate for $k(u)$ when

$$(7) \qquad 0 < \lambda < \tfrac{1}{2} \text{ and } \cos\pi\lambda \leq \nu < 1.$$

As one application of (2), we have proved

<u>Theorem 1</u>. The inequality of (6c),

$$k(u) \geq -\pi\lambda \sin(\pi-\beta)\lambda \ ,$$

remains valid in the range (7), with $\beta$ defined as in (6c).

Well known examples show equality is possible in (6c) for each $\lambda$ and $\nu$ satisfying (7), cf. [11, p. 117].

A similar improvement of (6a), when $\lambda > 1$ and $\nu$ is taken into account, seems very difficult, even for the subharmonic case $\nu = 0$ (a conjecture is offered in [8]).

In general, $k(u)$ depends on the relative sizes of $T(r,u)$, $N(r,u_2)$ and $N(r,u_1)$. There is a generalization of (2) which yields a bound for $k(u)$ in terms of $\lambda$, $\nu$ and $\nu_1 = \lim \sup N(r,u_1)/T(r,u)$, when $\lambda < 1$; work on this extension is still in progress.

A closely-related problem involves finding lower bounds for the "$\alpha$-spread"

$$(8) \qquad \sigma_\alpha(r,u) = \text{measure } \{\theta\varepsilon(-\pi,\pi): u(re^{i\theta}) > \alpha T(r)\},$$

$-\infty < \alpha < \infty$. This problem is completely solved in the most important case ($\alpha = 0$) by Baernstein [2]; and for $0 < \alpha < \infty$, by Anderson and Baernstein [1]. The method of our Theorem 1 gives new information when $\alpha < 0$ and $\lambda < 1$.

A proof of (2) and Theorem 1, and of the related results for $\alpha$-spreads, will be given in another paper. Here we consider some direct consequences of (2).

The previous proofs of each of (6a) - (6c) have been distinct; (2) permits a unified treatment of these as well as of Theorem 1.

1. __A proof of__ (6b). We compare here two new proofs of (6b). For this problem, take $a = o$ in (2), to get

$$(9) \qquad T^*(re^{i\theta}) \leq \int_o^R A(t,u) K_1(\tfrac{r}{t}, \theta)\frac{dt}{t} \;+$$

$$\qquad + \int_o^R N(t,u_2) K_2(\tfrac{r}{t}, \theta)\frac{dt}{t} + c_o\, T(2R)\, (\tfrac{r}{R})^{\frac{1}{2}} \qquad (0 < r < R).$$

We assume $T$ has Polya peaks $\{r_n\}$ of order $\mu < \tfrac{1}{2}$, i.e.

$$(10) \qquad T(r) \leq T(r_n)(r/r_n)^\mu(1 + \eta_n) \qquad (\eta_n r_n \leq r \leq r_n/\eta_n)$$

holds with $r_n \to \infty$ and $\eta_n \downarrow 0$, $\eta_n r_n \to \infty$. We put

$$(11) \qquad \beta_n = \tfrac{1}{2}\, \sigma_o(r_n,\, u)$$

(recall definition (8) above), and let

$$(12) \qquad \beta_* = \lim_{\substack{n\to\infty \\ n\varepsilon I}} \beta_n$$

denote any subsequential limit of $\{\beta_n\}$.

We assume now that $\nu < 1$; then it is elementary that $\beta_* > 0$ (in fact, by [6] and [2], necessarily $\beta_* \geq \min\{\pi,\ \cos^{-1}(\nu)/\mu\}$). Let

$$(13) \qquad k' = \max\{0, k(u)\}$$

where $k(u)$ was defined in (3). By our definitions (1), (8), (11) and (4), the choice $\theta = \beta_n$, $R = r_n/2\eta_n$ in (9) gives

M. ESSÉN, J. ROSSI, AND D. SHEA

$$T(r_n) \leq \int_{\eta_n r_n}^{r_n/2\eta_n} T(t) \{k'K_1(\frac{r_n}{t}, \beta_n) + \nu K_2(\frac{r_n}{t}, \beta_n)\} \frac{dt}{t} + o(T(r_n)) \quad (n \to \infty).$$

Using (10) and letting $n \to \infty$ through the subsequence I of (12), we deduce

$$1 \leq k'\hat{K}_1(\mu, \beta_*) + \nu\hat{K}_2(\mu, \beta_*)$$

where

$$\hat{K}_j(\mu, \theta) = \int_o^\infty K_j(t, \theta) t^{-\mu-1} dt.$$

Using the evaluations

$$(14) \qquad \hat{K}_1(\alpha, \theta) = \frac{\sin\theta\alpha}{\pi\alpha\cos\pi\alpha}, \quad \hat{K}_2(\alpha, \theta) = \frac{\cos(\pi-\theta)\alpha}{\cos\pi\alpha} \qquad (|\alpha| < \frac{1}{2})$$

we obtain

$$(15) \qquad \cos\pi\mu \leq k' \frac{\sin\beta_*\mu}{\pi\mu} + \nu\cos(\pi-\beta_*)\mu .$$

Since $k' \geq 0$ and $0 \leq \mu < \frac{1}{2}$, (15) still holds if we replace $\beta_*$ by $\pi$. Then (15) only gives information if $\nu < \cos\pi\mu$; in that case, $k' > 0$ and we deduce

$$(16) \qquad (\cos\pi\mu - \nu) \frac{\pi\mu}{\sin\pi\mu} \leq k(u).$$

Inequality (16) is strongest if we choose

$$(17) \qquad \mu = \inf \{\rho \geq 0: \liminf_{t,r \to \infty} \frac{T(tr)}{T(t)r^\rho} < \infty\} ,$$

the minimum choice for which $T(r)$ has peaks of order $\mu$ ([4]). Since $\mu \leq \lambda$, by (5) and (17), (6b) is proved.

The original proofs of (6b) for $u = \log|f|$ exploited the somewhat extraneous fact, due to Littlewood and Ahlfors, that there exist (many) complex c such that $w(z) = \log|f(z) - c| = w_1(z) - w_2(z)$ satisfies

$$(18a) \qquad N(r,w_1) \sim T(r,w) \sim T(r,u) \qquad (r \to \infty)$$

and, obviously, $w_2 = u_2$ and

$$(18b) \qquad A(r,w)^+ = A(r,u)^+ + O(1).$$

Although the proof of (16) just given avoids this procedure, we observe that a simpler device works well for general $\delta$-subharmonic functions. One merely replaces $u(z)$ by

$$w(z) = \max\{u_1(z), u_2(z)\} - u_2(z) = w_1(z) - w_2(z)$$

to achieve (18) in the sharp form: $w_2 = u_2$,

$$(18') \qquad N(r,w_1) = T(r,w) = T(r,u), \quad A(r,w) = A(r,u)^+.$$

(This remark eliminates an extraneous hypothesis in [15, Theorems 2 and 3].)

2. <u>Another convolution inequality</u>. If we use (18'), then we can replace (9) by a convolution inequality having the simpler form

$$(19) \quad T(r,u) \leq \int_0^R A(t,u)^+ Q_\nu\left(\frac{r}{t}\right)\frac{dt}{t} + E_\nu(r,R) \qquad (0 < r < R).$$

Here u is any $\delta$-subharmonic function with $u(0) = 0$ and

$$(20) \qquad 0 \leq \mu < \frac{1}{2}, \qquad \nu < \cos\pi\mu \ ;$$

$\nu$ and $\mu$ are defined in (4) and (17). The kernel $Q_\nu$ is positive, with

$$(21) \qquad \hat{Q}_\nu(\alpha) = \frac{\sin\pi\alpha}{\pi\alpha}\,(\cos\pi\alpha - \nu)^{-1}$$

for $|\alpha| < \rho$, where $\cos\pi\rho = \nu$, and

$$(22) \qquad E_\nu(r,R) = \left(\frac{c_o}{\cos\pi\sigma - \nu}\right)T(2R)\left(\frac{r}{R}\right)^\sigma + \eta(r)T(r) + \int_r^R \eta(t)T(t)K\left(\frac{r}{t}\right)\frac{dt}{t}$$

where K is a positive kernel with $\hat{K}(\alpha)$ convergent for $|\alpha| < \rho$, $\eta(r) \to 0$ when $r \to \infty$, $\sigma \in (o,\rho)$ is arbitrary. The derivation of (19) from (9) is given below.

Convolution inequalities like (19) are useful in studying extremal problems (compare [5]) and for obtaining results on the densities of sets where certain function-theoretic inequalities hold (cf. [9]). We shall use (19) to sharpen (6b) with the somewhat more informative

<u>Theorem 2</u>. If $u(z)$ satisfies (20), then

$$\liminf_{n \to \infty} \frac{A(r_n,u)}{T(r_n)} \geq \frac{\pi\mu}{\sin\pi\mu}\,(\cos\pi\mu - \nu)$$

holds for any sequence $\{r_n\}$ of strong peaks of order $\mu$ for $A(r,u)$.

The terminology just used means

$$(23) \qquad A(r) \leq A(r_n)\left(\frac{r}{r_n}\right)^\mu (1+\eta_n), \qquad T(r) \leq B_o\,A(r_n)\left(\frac{r}{r_n}\right)^\mu$$

both hold in $\eta_n r_n \leq r \leq r_n/\eta_n$ for a constant $B_o > 0$ and $\eta_n \to 0$, $\eta_n r_n \to \infty$. The fact that such sequences exist is itself a consequence of the convolution inequality (19), cf. Lemma 1.2 in [12].

The proof of Theorem 2 is immediate from (19) and (21) - (23).

To establish (19), we apply (9) to $w = u_1 \nu u_2 - u_2$; taking $\theta = \pi$ and using the properties (18'), we deduce

$$(24) \qquad T(r,u) \leq \int_o^R A(t,u)^+ K_1\left(\frac{r}{t},\pi\right)\frac{dt}{t} + \int_o^R N(t,u_2)K_2\left(\frac{r}{t},\pi\right)\frac{dt}{t} +$$

$$+ C_o T(2R)\left(\frac{r}{R}\right)^\sigma \qquad (0 < r < R),$$

where $\sigma(=\frac{1}{2})$ will be decreased below. We define $K_j(r) = K_j(r,\pi)$ and

$$(25) \qquad k(r) = \nu K_2(r) = \left(\frac{\nu}{\pi}\right)\frac{r^{1/2}}{1+r}\ , \qquad \hat{k}(\alpha) = \frac{\nu}{\cos\pi\alpha}\ .$$

In particular, for $0 \leq \alpha < \rho$, $\hat{k}(\alpha) < 1$ and we can define the positive "resolvent" kernel

$$(26) \qquad J = k + k*k + k*k*k + \ldots, \qquad \hat{J}(\alpha) = \frac{\nu}{\cos\pi\alpha - \nu}$$

where $k*k(r) = \int_0^\infty k(r/t)k(t)\,dt/t$.

Then (24) implies, in $0 < x < R$,

$$(27) \qquad \int_0^R T(r)J\left(\frac{x}{r}\right)\frac{dr}{r} \leq \int_0^R A(t,u)^+ (K_1 *J)\left(\frac{x}{t}\right)\frac{dt}{t} +$$
$$+ \int_0^R N(t,u_2)(K_2 *J)\left(\frac{x}{t}\right)\frac{dt}{t} + c_o \hat{J}(\sigma)T(2R)\left(\frac{x}{R}\right)^\sigma .$$

This inequality is useful only for $\sigma\epsilon(o, \rho)$, by (26).

Adding (27) to (24) yields

$$T(r) \leq \int_0^R A(t,u)^+ (K_1 + K_1 *J)\left(\frac{r}{t}\right)\frac{dt}{t} + \int_0^R N(t,u_2)(K_2 + K_2 *J)\left(\frac{r}{t}\right)\frac{dt}{t} -$$
$$- \int_0^R T(t)J\left(\frac{r}{t}\right)\frac{dt}{t} + c_o (1+\hat{J}(\sigma))T(2R)\left(\frac{r}{R}\right)^\sigma \qquad (o < r < R).$$

From (25) and (26),

$$J = k+k*J = \nu(K_2 + K_2 *J).$$

Thus, if we define

$$Q_\nu = K_1 + K_1 *J, \qquad K = K_2 + K_2 *J,$$

we have (19) and (22), with $\eta(r) = \max(\eta_1(r), \eta_2(r))$ and

$$\eta_1(r) = [N(r, u_2)/T(r) - \nu]^+ ,$$

$$\eta_2 = \eta_1 *K.$$

The evaluation (21) follows from (26) and (14).

## REFERENCES


[1]   J.M. Anderson and A. Baernstein II, The size of the set on which a meromorphic function is large, Proc. London Math. Soc. (3) 36 (1978), 518-539.

[2]   A. Baernstein II, Proof of Edrei's spread conjecture, Proc. London Math. Soc. (3) 26 (1973), 418-434.

[3]   _____________, Integral means, univalent functions and circular symmetrization, Acta. Math. 133 (1974), 139-169.

[4]   D. Drasin and D.F. Shea, Polya peaks and the oscillation of positive functions, Proc. Amer. Math. Soc. 34 (1972), 403-411.

[5]   _____________, Convolution inequalities, regular variation and exceptional sets, J. Analyse Math. 29 (1976), 232-293.

[6]   A. Edrei, Sums of deficiencies of meromorphic functions, J. Analyse Math. 14 (1965), 79-107.

[7]   _____________, A local form of the Phragmén - Lindelöf indicator, Mathematika 17 (1970), 149-172.

[8]   M. Essén and D.F. Shea, Applications of Denjoy integral inequalities to growth problems for subharmonic and meromorphic functions, Proc. Conf. on Classical Function Theory, Canterbury 1973, London Math. Soc. Lecture Notes, Series 12, 59-68.

[9]   _____________, Applications of Denjoy integral inequalities and differential inequalities to growth problems for subharmonic and meromorphic functions, Proc. Royal Irish Acad. 82 A (1982), 201-216.

[10] W.H.J. Fuchs, A theorem on min $\log|f(z)|/T(r,f)$, Proc. Conf. on Classical Function Theory, Canterbury 1973, London Math. Soc. Lecture Notes, Series 12, 69-72.

[11] W.K. Hayman, Meromorphic functions, Clarendon Press, Oxford, 1964.

[12] J. Miles and D.F. Shea, On the growth of meromorphic functions having at least one deficient value, Duke Math. J. 43 (1976), 171-186.

[13] I.V. Ostrovskii, Deficiencies of meromorphic functions of order less than one, Dokl. Akad. Nauk SSSR 150 (1963), 32-35.

[14] V.P. Petrenko, The growth of meromorphic functions of finite lower order, Izv. Akad. Nauk SSSR 33 (1969), 414-454.

[15] H. Ueda, On the minimum modulus of a subharmonic or algebroid function of $\mu_* < 1/2$, Kodai Math. J. 4 (1981), 298-314.



DEPARTMENT OF MATHEMATICS, UPPSALA UNIVERSITY
UPPSALA, SWEDEN

DEPARTMENT OF MATHEMATICS, VIRGINIA POLYTECHNIC UNIVERSITY
BLACKSBURG, VIRGINIA  24061

DEPARTMENT OF MATHEMATICS, UNIVERSITY OF WISCONSIN
MADISON, WISCONSIN  53706

Contemporary Mathematics
Volume **25**, 1983

# GEOMETRIC CONDITIONS FOR UNICITY OF HOLOMORPHIC CURVES

by

Leonard M. Smiley

The first unicity theorems in value distribution theory were proven by R. Nevanlinna in [6], including the following Five Values Theorem.  Let f and g be entire meromorphic functions and let $a_1, \ldots, a_5$ be distinct points (values) in $P^1$.  Under the condition that $(f|E)^{-1}(a_i) = (g|E)^{-1}(a_i)$, $i = 1, \ldots, 5$, where $E = C - \{\text{some finite disc}\}$, Nevanlinna proved $f \equiv g$.  This may be called an absolute unicity theorem inasmuch as the condition concerns set equality.  Relative unicity theorems add the requirement that, for each inverse image point in question, f and g take their common value there with the same multiplicity.  In [6] and thereafter the methods used in proving the relative theorems have been essentially different from those in the absolute case.

Recently, Hirotaka Fujimoto has extended the relative theory to the case of meromorphic maps from $C^m$ to $P^n$ in [4].  He has asked for a corresponding extension of the absolute theory [5].  A very general such extension is possible [7], using the Second Main Theorem of W. Stoll and P.M. Wong.  In this paper we give this extension in the case of holomorphic curves in which case new conditions of a geometric character appear (Thm 4.1).

Mention should be made of the work of S.J. Drouilhet [3] who gave the first several variable extensions of absolute unicity results.

## 1) Nevanlinna theory

The First and Second Main Theorems of Nevanlinna theory we use can be found at the needed level of generality in Cowen and Griffiths [2] so we use it as a self-contained reference.

By a <u>holomorphic curve</u> we mean a holomorphic mapping f: $C \to M$, where M is a compact, complex manifold.  If M is provided with a positive line bundle L and corresponding positively curved metric, we define the <u>characteristic function</u>

$$T(L,f,r) = \int_s^r \left( \int_{\Delta p} f^*\omega \right) \frac{dp}{p}$$

where $\omega$ is the (1,1) form associated to the metric, and $\Delta p$ is the p-disc.  We will consider two cases:  (1) $M = P^n$, $L = H$ the degree one bundle with Fubini-Study metric.  In this case it is well known that $T(L,r)$ is equivalent for asymptotic growth comparison with the original characteristic functions of Nevanlinna and Cartan, (2) $M = P^n \times P^n$, $L = \pi_1^* H \otimes \pi_2^* H := H'$ with the induced metric.

If D is a divisor in C, then as usual we have

$$n(D,r) = \text{degree of } (D \cap \Delta_r)$$

$$N(D,r) = \int_s^r n(D,p) \, \frac{dp}{p} \; .$$

If A is a divisor on M, $A_f := f^{-1}(A)$

$$N(A,f,r) = N(A_f,r).$$

Note:   The lower limit s in our integrals is assumed greater than zero, and for purely asymptotic results may be fixed arbitrarily large.  We suppress it in our notation.

If $A \in |L|$, and $f(C) \not\subset A$, we have [2, p. 111]

(1.1) $N(A,f,r) \leq T(L,f,r) + 0(1,A)$.

The system of divisors $|H|$ in $P^n$ is parametrized by the dual space $P^{n*}$. A set of such hyperplanes $A_1$, ..., $A_q$ is in <u>general</u> <u>position</u> if each $(n + 1)$-fold intersection of them is empty.  A holomorphic curve $f: C \to P^n$ may be given

$$f(\zeta) = [Z_0(\zeta), \ldots, Z_n(\zeta)]$$

giving also a vector in $C^{n+1}$

$$Z = (Z_0, \ldots, Z_n).$$

The Wronskian function

$$(Z \wedge Z' \ldots \wedge Z^{(n)})(\zeta) \text{ has, if it is not the zero function, a}$$

zero-divisor depending only on f.  We call this Wronski-divisor $D_n$.  In this case, we say f is <u>non-degenerate.</u>

(1.2) If f is a non-degenerate holomorphic curve in $P^n$, and $A_1$, ..., $A_q$ are hyperplanes in general position, then, for $\varepsilon > 0$

$$\sum_{j=1}^q N(A_j,f,r) \geq N(D_n,r) + (q - n - 1 - \varepsilon) \, T(H,r) + C \, ||$$

(We have substituted $N(D_n,r)$ for its formal equivalent in [2].

2) <u>The Product Curve</u>

In the sequel, we assume (1) f and g are distinct, non-degenerate holomorphic curves in $P^n$, $n > 1$ (2) $A_1$, ..., $A_q$ are $q > n$ hyperplanes of $P^n$ in general position (3) supp $(A_{i,f})$ = supp $(A_{i,g})$, $i = 1$, ..., q (4) $z \in$ supp

$(A_{i,f})$ implies $f(z) = g(z)$, $i = 1, \ldots, q$.

We now consider $h: = f \times g: C \to P^n \times P^n$. There is a natural subspace of $|H'|$ generated by sections of the form $\lambda_i \otimes \lambda_j - \lambda_j \otimes \lambda_i$ where the $\lambda$'s are hyperplane sections. The base-point set of this subspace is $\Delta_{P^n \times P^n}$, the diagonal. We call this subspace $||H'||$. By assumption (1), $f \times g(C)$ is not contained in the diagonal. Take $B \in ||H'||$ such that $f \times g(C) \not\subseteq B$.

__Lemma 2.1__  $N(B,h,r) \leq T(H,f,r) + T(H,g,r) + 0(1,B)$

__Proof__ Calculating the induced metric in $H'$ yields:

$$T(H',h,r) = T(H,f,r) + T(H,g,r) \text{ and } (1.1) \text{ applies. q.e.d.}$$

Define a divisor $S$ in $C$ by $S \leq 1$, supp $S = $ supp $(A_{i,f})$.

__Lemma 2.2__  $N(S,r) \leq N(B,h,r)$

__Proof__ $h(S) \subseteq \Delta_{P^n \times P^n}$, the diagonal, by assumption (4) above. Also, $\Delta \subseteq B$, so $S \subseteq h^{-1}(B)$. Comparing multiplicities gives the result. q.e.d.

We now define two numbers

(5)  $\ell = \min[n, \lim\sup_{|z| \to \infty} \{\max_i (\max(A_{i,f}(z), A_{i,g}(z)))\}]$

Let $T[i,q] = \{j \mid j: \{1, \ldots, i\} \to \{1, \ldots, q\}, \text{ injective}\}$

(6)  $d = \lim\sup_{|z| \to \infty} (\max \{i \mid \exists \, j \in T[i,q], z \in \text{supp } A_{j(k),f}, k = 1, \ldots, i\})$

__Remarks__ The number $\ell$ will be less than $n$ only if, outside a finite disc, neither $f$ nor $g$ intersect an $A_i$ to multiplicity $\geq n$. The number $d$ is at most $n$ by general position. If, outside any finite disc, $f$ (hence $g$) always hits $m$ - fold intersections of the $A_i$, then $d \geq m$.

## 3)  Divisor Comparisons

We introduce a "cut-off" counting function for $f$ (or $g$) with respect to a hyperplane $A$.

$$\bar{N}(A,f,r) = N(\min(A_f,n),r)$$

__Lemma 3.1__

$$\sum_{j=1}^{q} N(A_j,f,r) - N(D_{n,f},r) \leq \sum_{j=1}^{q} \bar{N}(A_j,f,r)$$

__Proof__

The statement is true if, for any point $x_0$ in supp$(A_{j,f})$ for any $j$, the inequality $\sum_{j=1}^{q} A_{j,f}(x_0) - D_{n,f}(x_0) \leq \sum_{j=1}^{q} \min(A_{j,f},n)(x_0)$ holds. We may assume that $f(x_0)$ lies on $A_1, \ldots, A_p$ only, with $p \leq n < q$. We may choose homogeneous coordinates $W_1, \ldots, W_{n+1}$ so that $W_j$ is the linear function defining $A_j$. If $\mu_j$ is the multiplicity of f with respect to $A_j$ at $x_0$ (i.e. $\mu_j = A_{j,f}(x_0)$), and $\zeta$ is a local coordinate around $x_0$, then the Wronskian function for f is given near $x_0$ by a determinant of the form

$$\left|
\begin{array}{cccc:ccc}
\zeta^{\mu_1}[\;] & \zeta^{\mu_2}[\;] & \cdots & \zeta^{\mu_p}[\;] & Z_{p+1} & \cdots & Z_{n+1} \\
\zeta^{\mu_1-1}[\;] & \zeta^{\mu_2-1}[\;] & \cdots & \zeta^{\mu_p-1}[\;] & Z'_{p+1} & & \\
\cdot & & & & & & \\
\cdot & & & & & & \\
\cdot & & & & & & \\
\zeta^{\mu_1-n}[\;] & & & \zeta^{\mu_p-n}[\;] & & &
\end{array}
\right|$$

where brackets represent holomorphic factors.

By reordering if necessary we may assume that $\mu_i \geq n$ only for $1 \leq i \leq p_1 \leq p$. Then $D_{n,f}(X_0) \geq (\mu_1 - n) + \ldots + (\mu_{p_1} - n)$.

By definition,

$$\sum_{j=1}^{q} A_{j,f}(x_0) = \mu_1 + \ldots + \mu_p.$$

So

$$\sum_{j=1}^{q} A_{j,f}(X_0) - D_{n,f}(x_0) \leq p_1 n + \mu_{p_1+1} + \ldots + \mu_p$$

$$= \sum_{j=1}^{q} \min(A_{j,f}, n)(x_0) \quad \text{q.e.d.}$$

<u>Lemma 3.2</u>  We may choose an $s > 0$ as the lower limit of integration such that

$$\sum_{j=1}^{q} \bar{N}(A_j, f, r) \leq \ell d N(S, r).$$

<u>Proof</u>:  This follows, on the divisor level, immediately from the definitions.

4)  <u>Unicity Results</u>

    <u>Theorem 4.1</u>

    Under assumptions (1) - (6),

$$q \leq n + 1 + 2\ell d$$

<u>Proof</u>:  Adding the inequalities (1.2) for f and g, we get, by Lemma 3.1

$$(q - n - 1)[T(H,f,r) + T(H,g,r)] \leq \sum_{j=1}^{q} \bar{N}(A_j,f,r) + \bar{N}(A_j,g,r) +$$

$$\varepsilon[T(H,f,r) + T(H,g,r)] + C \;\;||$$

Since $f \not\equiv g$, we can find $B \in |H'|$ as above, such that

$$\sum_{j=1}^{q} \bar{N}(A_j,f,r) + \bar{N}(A_j,g,r) \leq 2\ell d N(S,r)$$

$$\leq 2\ell d N(B,h,r)$$

$$\leq 2\ell d[T(H,f,r) + T(H,g,r)] + C$$

by Lemmas 2.1 and 2.2

Dividing by $[T(H,f,r) + T(H,g,r)]$ and letting $\varepsilon \to 0$, while taking $\lim \sup_{r \to \infty}$,

we get the result                                    q.e.d.

<u>Corollary 4.2</u>

Under assumptions (1) - (4), $q \leq 2n^2 + n + 1$

<u>Proof</u>

The numbers $\ell$ and $d$ are at most n.     q.e.d.

<u>Corollary 4.3</u>

Assume (1) - (4) and either

(i)  outside some finite radius, neither f nor g intersects the intersection of any two $A_i$'s.

<u>or</u>

(ii)  outside some finite radius, both f and g intersect all $A_i$'s <u>transversally.</u>

Then $q \leq 3n + 1$.

<u>Corollary 4.4</u>

Assume (1) - (4) and (i), (ii).  Then $q \leq n + 3$.

<u>Proof</u>

(i) $\to$ $d = 1$, (ii) $\to$ $\ell = 1$.                q.e.d.

<u>Remarks</u>   Note that the bounds in all three corollaries are generalizations of (4) for n = 1.  Of course the assumption (ii) brings us into "relative" unicity and this is the realm of Hirotaka Fujimoto.

In a forthcoming paper we deal with meromorphic maps to $P^n$ having domains of a class including all affine algebraic varieties, and treat linear dependence of a set of such maps, rather than identity of only two.  This allows us to generalize some classical results of H. Cartan [1].   The Second Main Theorem used is from [8].

L.M. SMILEY

## References


[1] H. Cartan, *Sur Les zéros des combinaisons linéaires de p fonctions holomorphes données*, Mathematica (Cluj) 7 (1933), 5-31.

[2] M. Cowen and Ph. Griffiths, Holomorphic curves and metrics of negative curvature, J. Analyse Math. 29 (1976), 93-152.

[3] S.J. Drouilhet, A unicity theorem for meromorphic mappings between algebraic varieties, Trans. Amer. Math. Soc. 265 (1981), no. 2, 349-358.

[4] H. Fujimoto, The uniqueness problem of meromorphic maps into the complex projective space, Nagoya Math. 58 (1975), 1-23.

[5] Kataka Conference, Open problems in geometric function theory, Proceed. 5 Int. Symp. Division Math., The Taniguchi Foundation, (available from: Dept. of Math., Osaka Univ., Toyonaka, 560 Japan), 1978, 22 pp.

[6] R. Nevanlinna, Einige Eindentigkeitssätze in der Theorie der meromorphen Funktionen, Acta. Math. 48 (1926), 367-391.

[7] L. Smiley, Dependence theorems for meromorphic maps (1979 Notre Dame Thesis), 57 pp.

[8] P.-M. Wong, Defect relations for meromorphic maps on parabolic manifolds (to appear in Duke Journal).



DEPARTMENT OF MATHEMATICS<br>
UNIVERSITY OF ALASKA<br>
ANCHORAGE, ALASKA 99508

Contemporary Mathematics
Volume **25**, 1983

## ON PSEUDO-PRIMALITY OF THE COMBINATION OF MEROMORPHIC FUNCTIONS
## SATISFYING LINEAR DIFFERENTIAL EQUATIONS

Guo-Dong Song        Chung-Chun Yang

## 1. Introduction

A meromorphic function $F(z)$ is said to be prime (psuedo-prime) if every factorization of the form $F=f(g)$ implies that either f is fractional linear or g is linear (either f is rational or g is polynomial). A transcendental meromorphic function $F(z)$ is said to be left prime, if every factorization of the above form implies that f is fractional linear whenever g is transcendental.

It has been an interesting problem to find the classes of prime or pseudo-prime entire and mermorphic functions. For instance, Prokopovich [5] discussed the primality of functions of the form

$$F(z) = \sum_{j=1}^{m} Q_j(z)e^{P_j(z)}, \quad F(z) \neq C+Q_2(z)e^{P_2(z)}, \qquad (1)$$

where $P_j$, $Q_j$ are polynomials, and C a constant. In a previous paper [6], it extended the above case by assuming $Q_j$ being rational functions. In particular, the following three results are obtained

**Proposition 1.** Suppose that $P_j(z)$ are polynomials with $n_m = \deg P_m > \deg P_j = n_j$, $j=1, \ldots, m-1 (m \geq 2)$, and that $Q_j(z)$ are rational functions with $Q_m(z) \not\equiv 0$. Then the function $F(z)$ of the form (1) is left prime.

**Proposition 2.** Suppose that in addition to the hypothesis of proposition 1 the function $F(z)$ given by (1) satisfies $0 \leq n_1 \leq n_2 < \ldots < n_m$. Then $F(z)$ is prime, unless all $P_j$ and $Q_j$ have a common right factor, that is, there exists a non-linear polynomial $g(z)$ such that

$$P_j = \bar{P}_j(g), \quad Q_j = q_j(g), \quad j=1,\ldots,m,$$

where $\bar{P}_j$ is a polynomial, $q_j$ is a rational function.

**Proposition 3.** Suppose that

$$h(z) = C+Q(z)e^{P(z)}$$

where C is constant, $Q(z) \not\equiv 0$ is rational, and $P(z)$ is a polynomial of degree $\geq 1$. Then $h(z)$ is prime, unless

(i)   $Q(z) = \beta(z)^n$ with $\beta(z)$ being rational and $n \geq 2$;

(ii) P and Q have a common right factor.

In this note, we shall deal with the pseudo-primality of a more general class of meromorphic functions. We shall use the following elegant result due to Steinmetz [7].

Theorem A.  Let $h(z)$ be the trascendental meromorphic function satisfying a linear differential equation

$$W^{(n)} + a_{n-1}(z) W^{(n-1)} + \ldots + a_o(z) W + a(z) = 0, \tag{2}$$

where $a(z)$, $a_o(z), \ldots, a_{n-1}(z)$ are rational functions. Then $h(z)$ is pseudo-prime.

We shall freely use the notations of Nevanlinna theory such as $T(r,f)$, $N(r,f)$, $m(r,f)$, $S(r,f)$, $\delta(o,f)$, etc. [2]. Our main results are the following:

Theorem 1.  Let D denote the class of mermorphic functions which satisfy a linear differential equation of the form (2). Let

$$F(z) = \sum_{j=1}^{m} Q_j(z) \phi_j(z),$$

where $\phi_j \in D$, and $Q_j$ is rational, $j=1,\ldots,m$. Then the function F and all its derivatives $F^{(n)}$, $n=1, 2,\ldots$, are pseudo-prime.

Theorem 2.  Suppose that in addition to the hypotheses of Theorem 1 the function $\phi_j$ satisfies

$$T(r,\phi_j) = o\{T(r,F)\} \text{ as } r \to \infty, \; j=1,\ldots,m-1 \; (m \geq 2)$$

and

$$N(r,\phi_m) + N(r,\frac{1}{\phi_m}) = o\{T(r,F)\} \text{ as } r \to \infty,$$

where $F(z)$ is of the form

$$F(z) \neq \sum_{j=1}^{m} Q_j(z)\phi_j(z), \; F(z) \neq Q(z) + Q_2(z)\phi_2(z) \tag{3}$$

with $Q_j$ being non-zero rational functions and Q being any polynomial. Then the function F and all its derivatives $F^{(n)}$, $n=1, 2,\ldots$, are left prime.

Remark.  This theorem gives an extention of proposition 1.


2.  Preliminaries.

Let M be the family of all meromorphic functions, and R the family of all rational functions.  Clearly, M is a linear space over R: i.e., if $\phi_1, \phi_2 \in$ M and $R_1 \in$ R, then $R_1\phi_1 + \phi_2 \in$ M.  Recall that the subset D⊂M is defined such that $\phi \in$ D if and only if $\phi$ satisfies the lienar differential equation (2), where n, a(z), and $a_j(z)$ all depend on $\phi$.  Also, for $\phi \in$ D, we define the subset $\phi$ of M in such a way: $L_\phi$ is spanned by 1, $\phi'$,..., $\phi^{(n)}$ over R, which is denoted by

$$L_\phi = \langle 1, \quad , \phi', \ldots, \phi^{(n)} \rangle. \tag{4}$$

The following properties concerning D and $L_\phi$ can be easily derived

$1^{\circ}$.  $\phi \in$ D, $R_1 \in$ R $\longrightarrow$ $R_1\phi \in$ D and $R_1 + \phi \in$ D

$2^{\circ}$.  $\phi \in$ D $\longrightarrow$ $L_\phi$ is a linear subspace of M over R;

$3^{\circ}$.  $\phi \in$ D $\longrightarrow$ $\phi^{(k)} \in L_\phi \cap$ D for every non-negative integer k.

<u>Lemma 1.</u>  The family D is a linear subspace of M over R.

<u>Proof.</u>  By property $1^{\circ}$, it suffices to show that if $\phi$, $\psi \in$ D, then $\phi + \psi \in$ D. Suppose that $\phi$ and $\psi$ satisfy; respectively equation (2) and the following linear differential equation

$$w^{(\ell)} + b_{\ell-1}(z) w^{(\ell-1)} + \ldots + b_o(z)\, w + b(z) = 0, \tag{5}$$

where b, $b_o, \ldots, b_{\ell-1} \in$ R.  Let $L_\phi$ be defined by (4) and

$$L_\psi = \langle , 1, \psi, \psi', \ldots, \psi^{(\ell)} \rangle.$$

Also, let $L = L \oplus L_\psi$, i.e.,

$$L = \langle 1, \quad , \phi', \ldots, \phi^{(n)}, \psi, \psi', \ldots, \psi^{(\ell)} \rangle. \tag{6}$$

From (2), (5) and (6) we can see that the dimension of L

$$\dim L \leq n + \ell + 1.$$

On the other hand, by property $3^{\circ}$, $(\phi+\psi)^{(k)} \in$ L for k = 0, 1, 2,... Hence, there exist c(z), $c_o(z), \ldots, c_{n+\ell}(z) \in$ R such that

$$(\phi+\psi)^{(n+\ell+1)} + c_{n+\ell}(\phi+\psi)^{(n+\ell)} + \ldots + c_o(\phi+\psi) + c = 0.$$

And this means $\phi + \psi \in$ D.                                    q.e.d.

We shall also need the following

Lemma 2.  Let $g(z)$ be a transcendental meromorphic function such that

$$N(r,g) = o\{T(r,g)\} \quad \text{as } r\to\infty.$$

Let $R(z)$ be a rational function of degree $k$ and $\alpha(z)$ ($\neq$ constant) a mermorphic function satisfying the condition $T(r,\alpha) = o\{T(r,g)\}$ as $r\to\infty$.  Then

$$\lim_{\substack{r\to\infty \\ r\notin E}} \frac{N(r, \frac{1}{R(g)-\alpha})}{T(r,g)} \geq k-1 + \delta(o,g), \tag{7}$$

where the measure of $E$ is finite.

### 3.  Proofs of Main Theorems

Proof of Theorem 1.  Since $\phi_j$ satisfies the linear differential

equation (2), $\phi_j \in D$.  Hence, by lemma 1, $F= \sum_{j=1}^{m} Q_j\phi_j \in D$.  Also by property $3^o$,

$F^{(k)} \in D$, $k=0,1,2,\ldots$, satisfies a linear differential equation with rational

functions as the coefficients.  Then by Theorem A, $F^{(k)}$ is pseudo-prime.

Proof of Theorem 2.  By Theorem 1, $F^{(k)}$ is pseudo-prime, $k=0,1,2\ldots$
We deal with the left primality of $F(z)$ first.  Suppose that $F=R(g)$, where $g$
is transcendental entire, and $R(z)$ is a rational function of degree $k$.
Apparently,

$$T(r,F) = O\{T(r,g)\} \quad (R\to\infty).$$

Let

$$\alpha(z) = \sum_{j=1}^{m-1} Q_j(z)\, \phi_j(z).$$

We have

$$T(r,\alpha) = o\{T(r,F)\} = o\{T(r,g)\} \quad \text{as } r\to\infty.$$

$$N(r_n, \frac{1}{\phi_m}) \geq (k-1 + o(1))T(r_n,g), \quad n\to\infty.$$

But this can hold only if $k=1$ by the assumption, namely, $R(z)$ is fractional
linear.

Consider now the left primality of $F'(z)$.  Suppose that $F'=R_1(g_1)$ with

$g_1$ being transcendental entire and $R_1$ rational.  We have

$$F'= \sum_{j=1}^{m=1} (Q_j'\phi_j + Q_j\phi_j') + (Q_m\phi_m)'$$

$$= \beta(z) + \psi'(z) \quad (\text{where } \psi=Q_m\phi_m.)$$

Clearly, $T(r,\beta) = o\{T(r,g_1)\}$. It suffices to sho that

$$N(r,\psi') + N(r, \frac{1}{\psi'}) = o\{T(r, F)\}. \tag{8}$$

Indeed, we have

$$N(r,\psi') \leq 2N(R,\psi) = o\{T(r,F)\}$$

and

$$N(r, \frac{1}{\psi'}) + m(r, \frac{1}{\psi}) \leq N(r, \frac{1}{\psi'}) + m(r, \frac{1}{\psi'}) + m(r, \frac{\psi'}{\psi}) + o(1)$$

$$\leq T(r,\psi') + S(r,\psi) \leq T(r,\psi) + N(r,\psi) + S(r,\psi)$$

$$\leq m(r, \frac{1}{\psi}) + N(r, \frac{1}{\psi}) + N(r,\psi) + S(r,\psi).$$

By a theorem of Wittich [8], $\psi$ is of positive and finite order, so that

$$S(r,\psi) = O(\log r) = o\{T(r,F)\} \text{ as } r \to \infty.$$

Thus we obtain (8).

Repeating the above procedure, we deduce that $F^{(k)}$, k=1,2,..., are all left prime.                                        q.e.d.

## 4.  Applications

In Theorem 1, each $\phi_j$ can be taken as one of the following functions: $\exp(p(z)), \sin(p(z))$, where $p(z)$ is a polynomial; the n-th Bessel function $J_n(z)$

(where $J_n(z)$ satisfies $w'' + \frac{1}{z} w' + (1 - \frac{2^n}{z^2}) w = 0)$.

Ozawa [3] proved the primality of the function

$$\int_z^O e^{-t^p} dt, \quad p(\geq 2) \text{ is an integer.}$$

He also pointed out that one can find several prime functions in the class of functions having the form

$$F(z) = \int_O^z Q(t)e^{P(t)} dt, \tag{9}$$

where P and Q are polynomials with $\deg Q \leq \deg P - 2$.  The following theorem gives a fairly complete answer to this type of problem.

Theorem 3.  The function F(z) of the form (9) is prime, unless there are polynomials $\alpha(z)$, $\underline{p}(z)$ and $q(z)$ such that

$$P(z) = \underline{p}(\alpha(z)), \qquad Q(z) = \alpha'(z)q(\alpha(z)), \tag{10}$$

where $\deg \alpha \geq 2$.

$\underline{\text{Proof.}}$  By Theorem 1, $F(z)$ is pseudo-prime.  Let $F=f(g)$, then

$$Qe^P = g'f'(g).  \tag{11}$$

We discuss two cases separately.

(i)     $g$ is a polynomial of degree $\geq 2$.  Then

$$\frac{Q}{g'} e^P = f'(g).$$

Hence, by proposition 3, we obtain (10) with $\alpha = g$.

(ii)     $f$ is a polynomial and $g$ is transcendental entire.  Then from (11) and by Picard theorem $f'$ has at most one zero.  If $f'$ has a zero, then $f'$ and $g$ must have the following forms, respectively,

$$f'(\xi) = K(\xi - a)^n, \quad g(z) = a + \phi(z)e^{\psi(z)}, \tag{12}$$

where $K \neq 0$ and $a$ are constants, and $n \geq 1; \phi$ and $\psi$ are polynomials with $\phi \not\equiv 0$ and $\deg \psi = \deg P$.  From (11) and (12), we have

$$Qe^P = K \phi^n(\phi' + \phi\psi')e^{(n+1)\psi},$$

Hence

$$Q = K_1(\phi^n\phi' + \phi^{n+1}\psi') \quad K_1 (\neq 0) \text{ is a constant}$$

Thus

$$\deg Q \geq \deg \psi' = \deg P - 1,$$

which violates the assumption of the theorem.  Therefore, we must have $\deg f' = 0$, i.e., $f$ is linear                          q.e.d.

There are several classes of entire functions which and all their derivatives are shown to be prime.  For instance, Yang and Niino [9] proved that the function

$$F(z) = \int_0^z \{H_1(z)e^z + H_2(z)\}e^{p(z)} \, dz$$

and all $F^{(n)}(z)$ are prime, where $H_1$ and $H_2$ are non-zero entire functions of order $<1$, and $p(z)$ is a polynomial of degree $\geq 3$ with $p(z+c) - p(-z+c)$ being non-linear for any constant $c$.  We shall give another example.

$\underline{\text{Theorem 4.}}$  Let $P, P_1, \ldots, P_m, Q_1, \ldots, Q_m$ be polynomials such that $Q_j \not\equiv 0$ for $j = 1, \ldots, m$ and $\deg P > \deg P_m \ldots > \deg P_1$.  Let

$$F(z) = \int_0^z [Q_1(t)e^{P_1(t)} + \ldots + Q_m(t)e^{P_m(t)}] \, dt + e^{P(z)}.$$

Then $F(z)$ is prime, unless there are polynomials, $\alpha$, $\bar{p}$, $\bar{p}_j$ and $q_j$ such that

$$P = \bar{p}(\alpha), \quad P_j = \bar{p}_j(\alpha) \text{ and } Q_j = \alpha'q_j(\alpha); \quad j=1,\ldots, m, \tag{13}$$

where deg $\alpha \geq 2$.

Furthermore, if $P(z)$ is prime, then all derivatives $F^{(n)}$, $n=0,1,2,\ldots$, are prime.

Proof. From Theorems 1 and 2, it is readily seen that $F(z)$ is left prime. Suppose that $F=f(g)$ with $g$ being a polynomial of degree $\geq 2$. Then

$$Q_1 e^{P_1} +\ldots+ Q_m e^{P_m} + P'e^{P} = g'f'(g).$$

Therefore, we conclude from proposition 2 that there exist polynomials $\underline{p}$, $\bar{p}_1,\ldots, \bar{p}_m$ $q_1,\ldots, q_m$, and $\bar{p}$ such that

$$P_j = p_j(g), \quad Q_j = g'q_j(g) \text{ for } j=1,\ldots, m$$

and

$$\frac{P'}{g'} = \bar{p}(g), \quad P = \underline{p}(g). \tag{14}$$

Therefore, we have $p' = g'\bar{p}(g)$ and $p' = g'\underline{p}'(g)$ (from the second equation in (14). Hence, $\bar{p} = \underline{p}'$, so (13) follows with $\alpha = g$.

If $P(z)$ is prime, then by proposition 2, it is clear that all $F^{(n)}$, $n=0$, $1,2,\ldots$, are prime.

q.e.d.

Remark. In the proofs of Theorems 3 and 4, we just proved the primality in entire sense, that is, in the factorizaton $F = f(g)$ only entire factors $f$ and $g$ are considered. To complete the proofs, we only need to make use of Gross' result, which claims that every non-periodic entire prime function in entire sense in prime [1].

We are going to show, for instance, the non-periodicity of $F$ in Theorem 3. It suffices to show that of $F'$. Suppose that $F' = Qe^{P}$ were periodic with a period $\tau \neq 0$. Then

$$Q(z+\tau)e^{P(z+\tau)} = Q(z)e^{P(z)},$$

which implies

$$\frac{Q(z)}{Q(z+\tau)} = e^{P(z+\tau)-P(z)}$$

But this is impossible, since it is easily seen that $\deg(P(z+\tau)-p(z)) > 1$ if deg $P \geq 2$.

Finally to the verification of non-periodicity of $F^{(n)}$ in Theorem 4, we need a more delicate consideration.  The details are omitted here.

Concluding Remark.  It seems that thus far no method has been developed to deal the primality of functions of the form $F = \sum_j Q_j(z) e^{P_j(z)}$, where among the $p_j(z)$ two or more of them having the highest degree.  For instance, $F(z) = e^{z^2} + z e^{2z^2}$.  For such a specific function we can prove that it is a prime function.  In general we pose the following:

Conjecture.  Let $F(z) = \sum_{j=1}^{m} Q_j e^{P_j}$, where $Q_j$ and $P_j$ are polynomials.

Furthermore, suppose that $P_j$ (j=1,2,..., m) are linear independent over rational numbers and one of the functions $Q_j$ or $P_j$ is a prime polynomial, then F is prime.

Acknowledgement.  The first author is deeply grateful to Professor W. Fuchs for his help and supervision.

REFERENCES


1.  F. Gross, "Factorization of Entire Functions Which are Periodic Mod g", Indian Journ. pure Appl. Math. 2 (1971), 561-571.

2.  W. K. Hayman, "Meromorphic Functions", Oxford press, 1964.

3.  M. Ozawa, "On Prime Entire Functions", Kodai Math. Sem. Rep., 22 (1970).

4.  G.S. Prokopovich, "On Fixed Points of Mermorphic Functions", Ukr. Mathem. Zh., 25, No. 2 (1973), 248-260.

5.  _____, "On Superposition of Some Entire Functions," Ekr. Matem. Zh., 2b, No. 2 (1974), 188-195.

6.  G.D. Song, "On Primality of the Combination of Exponential Functions", to appear.

7.  N. Steinmetz, "Uber die Faktorisierbaren Lösungen Gewöhnlicher Differentialgleichungen", Math. Zeit., 170 (1980), 169-180.

8.  H. Wittich, "Neuere Untersuchungen Über Eindeutize Analytische Funktionen.  Berlin-Heidelberg-New York, Springer, 1968.

9.  C.C. Yang and K. Niino, "On Prime Entire Functions", Kodai Math. J. 5 (1982), 84-88.



Department of Mathematics     Naval Research Laboratory
Cornell University       Washington, D.C.  20375
Ithaca, New York  14853

Contemporary Mathematics
Volume 25, 1983

# A VALUE-DISTRIBUTION CHARACTERIZATION

# OF THE SMIRNOV CLASS

Charles S. Stanton

## 1. Introduction.

Let f be analytic in the unit disc. If the means $\int_0^{2\pi} \log^+ |f(re^{i\Theta})| d\Theta$ are

uniformly bounded, $0 < r < 1$, f is said to belong to the Nevanlinna class N.

It is well known that in this case f has radial limits $F(e^{i\Theta})$ a.e. If

$$\lim_{r \to 1} \int_0^{2\pi} \log^+ |f(re^{i\Theta})| d\Theta = \int_0^{2\pi} \log^+ |F(e^{i\Theta})| d\Theta \, , \text{ f is said to belong to the}$$

Smirnov class $N^+$ (see [2] for details and the connection with factorization).

Let $N(r,w) = N_f(r,w)$ be the Nevanlinna counting function for the value w

in the complex plane. Jensen's formula gives, for $w \neq f(0)$,

$$N(r,w) = \frac{1}{2\pi} \int_0^{2\pi} \log |f(re^{i\Theta}) - w| d\Theta - \log |f(0) - w|. \tag{1}$$

We will also need the following classical formula of H. Cartan (see [4] p.8):

$$\int_0^{2\pi} \log^+ |f(re^{i\Theta})| d\Theta = \int_0^{2\pi} N(r,e^{i\phi}) d\phi + 2\pi \log^+ |f(0)|. \tag{2}$$

In Proposition 1, below, we give an extension of these formulas to the

boundary values $F(e^{i\Theta})$. From this we derive

__Theorem 1.__  $f \in N^+$ if and only if

$$\limsup_{\substack{r \to 1 \\ |w| \to \infty}} N(r,w) = 0.$$

Suppose D is an unbounded domain in C. The point $\infty$ is said to be regular

for the Dirichlet problem for D if D possesses a barrier at $\infty$ (see [6]). As

a corollary of Theorem 1, we derive the following result of P. Ahern and

W. Cohn [1]:

__Theorem A.__  Suppose f is analytic in the unit disc and maps into an unbounded

domain D. If $\infty$ is a regular point for D then f is in the Smirnov class $N^+$.

## 2. Properties of N(w) and its Riesz measure.

We recall some properties of the counting function N(r,w) (see [3] for

details). For each $r < 1$, $N(r,w) + \log |f(0) - w|$ is a subharmonic function of w.  Since $N(r,w)$ is increasing in r the limit $N(1,w) = \lim_{r \to 1} N(r,w)$ exists,

with $0 \leq N(1,w) \leq +\infty$ .  For functions in the Nevanlinna class $N(1,w)$ is uniformly bounded except near $f(0)$.  As an increasing limit of subharmonic functions, $N(1,w)$ need not be subharmonic, but its upper regularization, defined by

$$N(w) = \lim_{\zeta \to w} \sup N(1,\zeta),$$

will be subharmonic except for the pole at $f(0)$.  Moreover, $N(w) = N(1,w)$ except on a set of logarithmic capacity zero.

For our purposes, we wish to extend $N(w)$ to the extended plane by setting

$$N(\infty) = \lim_{|\zeta| \to \infty} \sup N(1,\zeta). \tag{3}$$

With this definition $N(w)$ is subharmonic on the extended plane excluding $f(0)$.

We denote the Riesz measures of $N(r,w) + \log|f(0) - w|$ and $N(w) + \log|f(0) - w|$ by $\nu_r$ and $\nu$ respectively.  Thus $\nu$ and $\nu_r$ are defined as linear functionals on $\Psi \epsilon C_0^\infty$ by

$$\int_C \Psi(w)d\nu_r(w) = \frac{1}{2\pi} \int_C (N(r,w) + \log|f(0) - w|)\Delta\Psi(w)dudv \tag{4}$$

and

$$\int_C \Psi(w)d\nu(w) = \frac{1}{2\pi} \int_C (N(w) + \log|f(0) - w|)\Delta\Psi(w)dudv. \tag{5}$$

The following "change of variables" theorem characterizes the measures $\nu$ and $\nu_r$:

Theorem 2.  Suppose $\Psi$ is a Borel measurable function on the complex plane. Then

$$\frac{1}{2\pi} \int_0^{2\pi} \Psi(F(e^{i\Theta}))d\Theta = \int_C \Psi(w)d\nu_r(w) \tag{6}$$

and

$$\frac{1}{2\pi} \int_0^{2\pi} \Psi(F(e^{i\Theta}))d\Theta = \int_C \Psi(w)d\nu(w). \tag{7}$$

Moreover, the measures $\nu_r \to \nu$ in the weak - $*$ topology.

Proof.  Define $f_r$ by $f_r(e^{i\Theta}) = f(re^{i\Theta})$.  The functions $f_r$ and F defined on the unit circle induce Borel measures $f_r^* d\Theta$ and $F^* d\Theta$ on the complex plane (see e.g. [5] p. 318) such that for any Borel measurable function $\Psi$

$$\int_0^{2\pi} \Psi(f_r(e^{i\Theta})\,d = \int_C \Psi(\zeta)f_r^*\,d\Theta(\zeta) \tag{8}$$

and

$$\int_0^{2\pi} \Psi(F(e^{i\Theta}))\,d\Theta = \int_C \Psi(\zeta)F^*\,d\Theta(\zeta). \tag{9}$$

In particular,

$$N(r,w) + \log|f(0) - w| = \frac{1}{2\pi} \int_0^{2\pi} \log|f(re^{i\Theta}) - w|\,d\Theta$$

$$= \frac{1}{2\pi} \int_C \log|\zeta - w|f_r^*\,d\Theta(\zeta).$$

This represents $N(r,w) + \log|f(0) - w|$ as a potential of the measure $f_r^*\,d\Theta$. By uniqueness of the measure $d\nu_r$, we conclude

$$d\nu_r = \frac{1}{2\pi}f_r^*\,d\Theta$$

and (6) holds.

Before proving (7), we show a week-$*$ convergence. Recall that $N(w) = N(1,w)$ except on a set of capacity (and hence Lebesque measure) zero. From this and the defintions ((4) and (5)) of $\nu_r$ and $\nu$ as linear functionals, we deduce for $\Psi\varepsilon C_0^\infty$

$$\lim_{r\to 1} \int_C \Psi(w)\,d\nu_r(w)$$

$$= \lim_{r\to 1} \frac{1}{2\pi} \int_C (N(r,w) + \log|f(0) - w|)\Delta\Psi(w)\,dudv$$

$$= \frac{1}{2\pi} \int_C (N(1,w) + \log|f(0) - w|)\Delta\Psi(w)\,dudv$$

$$= \frac{1}{2\pi} \int_C (N(w) + \log|f(0) - w|)\Delta\Psi(w)\,dudv$$

$$= \int_C \Psi(w)\,d\nu w),$$

i.e. $\nu_r \to \nu$ weak-$*$ .

To prove (7), we note that for $\Psi\varepsilon C_0^\infty$

166           C. STANTON

$$\frac{1}{2\pi} \int_0^{2\pi} \Psi(F(e^{i\Theta}))\,d\Theta = \lim_{r\to 1} \int_C \Psi(w)\,d\nu_r(w)$$

$$= \int_C \Psi(w)\,d\nu(w).$$

From (9) we conclude that $2\pi d^* = F^* d\Theta$, and hence (7) holds for all Borel measurable functions.

3. <u>Jensen and Cartan Formulas for boundary values.</u>

We now show that the "meromorphic" versions of Jensen's and Cartan's formulas hold for the boundary values $F(e^{i\Theta})$.

<u>Proposition.</u>  Suppose $f \varepsilon N$.  Then

$$N(w) = \frac{1}{2\pi} \int_0^{2\pi} \log|F(e^{i\Theta}) - w|\,d\Theta - \log|f(0) - w| + N(\infty), \tag{10}$$

and

$$\frac{1}{2\pi} \int_0^{2\pi} \log^+|F(e^{i\Theta})|\,d\Theta \tag{11}$$

$$= \frac{1}{2\pi} \int_0^{2\pi} N(e^{i\phi})\,d\phi + \log^+|f(0)| - N(\infty)$$

(recall that $N(\infty)$ is defined by (3)).

<u>Proof.</u>  We define a function on the plane by

$$h(w) = \int_C \log|\zeta - w|\,d\nu(\zeta) - \log|f(0) - w|. \tag{12}$$

By the change of variables (7),

$$h(w) = \frac{1}{2\pi} \int_0^{2\pi} \log|F(e^{i\Theta}) - w|\,d\Theta - \log|f(0) - w|. \tag{13}$$

Since f is of bounded characteristic, $h(w)$ is in fact a subharmonic function in C with a logarithmic pole at $f(0)$.  The functions $h(w)$ and $N(w)$ have the same Riesz measure and hence differ by a harmonic function defined on the entire plane.

If $|w| > \max(|f(0)|, 1)$, then

$$|\zeta - w|/|f(0) - w| \le |\zeta| + 2,$$ and hence $h(w) \le \int_C \log^+(|\zeta| + 2)\,d\nu(\zeta).$

Since $N(w) \geq 0$, we conclude that the harmonic function $h(w) - N(w)$ is bounded above and hence constant.

We denote this constant by $k$, and observe that for $0 < \rho < \infty$

$$k = \frac{1}{2\pi} \int_0^{2\pi} [h(\rho e^{i\phi}) - N(\rho e^{i\phi})] d\phi \tag{14}$$

Since $N(w)$ is subharmonic in a neighborhood of $\infty$, we can use upper semi-continuity and the sub-mean value property to conclude

$$\lim_{\rho \to \infty} \frac{1}{2\pi} \int_0^{2\pi} N(\rho e^{i\phi}) d\phi = N(\infty) .$$

We also have

$$\frac{1}{2\pi} \int_0^{2\pi} h(\rho e^{i\phi}) d\phi = \frac{1}{2\pi} \int_0^{2\pi} [\int_C \log|\zeta - \rho e^{i\phi}| d\nu(\zeta) - \log|f(0) - \rho e^{i\phi}|] d\phi$$

$$= \int_C \log^+|\frac{\zeta}{\rho}| d\nu(\zeta) - \log^+\frac{|f(0)|}{\rho} ,$$

and hence $\displaystyle\lim_{\rho \to \infty} \int_0^{2\pi} h(\rho e^{i\phi}) d\phi = 0$.

From (14), we conclude $k = -N(\infty)$, and thus $N(w) = h(w) + N(\infty)$. Substituting the expression (13) for $h(w)$ completes the proof of Jensen's formula (10).

To derive Cartan's formula (11), we write

$$\log^+|F(e^{i\Theta})| = \frac{1}{2\pi} \int_0^{2\pi} \log|F(e^{i\phi}) - e^{i\phi}| d\phi \quad \text{and employ the version of Jensen's}$$

formula we just proved. Thus,

$$\frac{1}{2\pi} \int_0^{2\pi} \log^+|F(e^{i\Theta})| d\Theta$$

$$= \frac{1}{2\pi} \int_0^{2\pi} \frac{1}{2\pi} \int_0^{2\pi} \log|F(e^{i\phi}) - e^{i\phi}| d\phi d\Theta$$

$$= \frac{1}{2\pi} \int_0^{2\pi} [N(e^{i\phi}) + \log|f(0) - e^{i\phi}| - N(\infty)] d\phi$$

$$= \frac{1}{2\pi} \int_0^{2\pi} N(e^{i\phi}) d\phi + \log^+|f(0)| - N(\infty) .$$

### 4. Proof of Theorem 1.

We now prove Theorem 1.  Since $N(r,w)$ is increasing in $r$,

$$\limsup_{\substack{r \to 1 \\ |w| \to \infty}} N(r,w) = \limsup_{|w| \to \infty} N(1,w) = N(\infty),$$

and so it suffices to show $f \in N^+$ if and only if $N(\infty) = 0$.

From the classical Cartan identity (2), we deduce

$$\lim_{r \to 1} \frac{1}{2\pi} \int_0^{2\pi} \log^+ |f(re^{i\theta})|\, d\theta$$

$$= \lim_{r \to 1} \frac{1}{2\pi} \int_0^{2\pi} N(r,e^{i\phi})\, d\phi + \log^+ |f(0)|$$

$$= \frac{1}{2\pi} \int_0^{2\pi} N(1,e^{i\phi})\, d\phi + \log^+ |f(0)|$$

$$= \frac{1}{2\pi} \int_0^{2\pi} N(e^{i\phi})\, d\phi + \log^+ |f(0)|.$$

The last equality follows since $N(1,e^{i\phi}) = N(e^{i\phi})$ except possibly on a set of capacity zero, and hence a.e. ($\phi$).

Comparing the above equality with the boundary value version (11), we see that

$$\lim_{r \to 1} \int_0^{2\pi} \log^+ |f(re^{i\theta})|\, d\theta = \int_0^{2\pi} \log^+ |F(e^{i\theta})|\, d\theta + N(\infty).$$

Hence $f \in N^+$ if and only if $N(\infty) = 0$, which finishes the proof.

We can now prove the Theorem of Ahern and Cohn.  Assume $f$ maps the unit disc into a domain D and $\infty$ is a regular point for D.  This implies the boundary of D has positive capacity and hence D has a generalized Green's function $G(w,f(0))$.

The Green's function tends to zero at each regular boundary point (see e.g. [6] p. 17).  Since $N(w)$ is subharmonic with a pole at $f(0)$, $N(w) \leq G(w,f(0))$, and hence $N(w)$ also goes to zero at regular boundary points.  Thus, $N(\infty) = 0$ and, by Theorem 1, $f \in N^+$.

We should note that Ahern and Cohn also prove that if every function mapping into an unbounded domain D is in the Smirnov class, then D possesses a barrier at $\infty$.

## References


1.  P. Ahern and W. Cohn, A geometric characterization of $N^+$ domains, preprint.

2.  P. Duren, Theory of $H^p$-spaces, Academic Press, 1970.

3.  M. Essen and D.F. Shea, On some questions of uniqueness in the theory of symmetrization, Ann. Acad. Sci. Fennicae series A.I. Math. 4 (1978/79), 311-340.

4.  W.K. Hayman, Meromorphic Functions, Oxford University Press, 1964.

5.  H.L. Royden, Real Analysis, second edition, MacMillan, 1964.

6.  M. Tsuji, Potential theory in modern function theory, Maruzen, Tokyo, 1959.



Department of Mathematics
University of North Carolina at Chapel Hill
Chapel Hill, N.C.  27514

Contemporary Mathematics
Volume **25**, 1983

## ASYMPTOTIC PROPERTIES OF ENTIRE TRANSCENDENTAL SOLUTIONS
## OF ALGEBRAIC DIFFERENTIAL EQUATIONS

### Sh. Strelitz

### §1.    Introduction

Suppose the Taylor series

$$f(z) = \sum_{n=0}^{\infty} a_n z^n \qquad (1.1)$$

to converge in the disc $|z| < R$.

It will be useful to recall the definition of the maximal term $\mu(r)$ and the
central index $\nu(r)$ of the power expansion (1.1).  The maximal term is defined as

$$\mu(r) = \max_{n} \, |a_n| r^n; \; r < R. \qquad (1.2)$$

It is well known [11], [12] that there is only a finite number of indices
$n_1 < n_2 < \ldots < n_p$ for which

$$|a_{n_j} r^{n_j} k| = \mu(r), \; j=1,2,\ldots,p.$$

The central index $\nu(r)$ of (1.1) is then defined equal $n_p$: $(r)=n_p$, so that

$$|a_{\nu(r)}| r^{\nu(r)} \geq |a_{\nu(r)-s}| r^{\nu(r)-s}, \; s=1,2,\ldots,\nu(r)-1$$

and

$$|a_{\nu(r)}| r^{\nu(r)} > |a_{\nu(r)+s}| r^{\nu(r)+s}, \; s=1,2,3\ldots \; .$$

We intende to define a generalized index.  For this purpose let

$$f(z) = \sum_{n=0}^{\infty} \frac{f^{(n)}(a)}{n!} (z-a)^n \qquad (1.3)$$

be the Taylor expansion (1.1) about the point a.  Let further $p(r)$ be an
increasing function on the segment $[r',r'']$ such that $\ln p(r)$ is a convex function of $\ln r$ on $[r',r'']$.  We define the maximal term of (1.3) with respect to
$p(r)$ for $r' \leq r \leq r''$ as

$$\mu_a(p(r)) = \max_{n} \, \left|\frac{f^{(n)}(a)}{n!}\right| p^n(r) \qquad (1.4)$$

(with the same a as in(1.3)) and the central index $\nu_a(r|p)$ as the maximal index
of those terms $|a_{n_j}| p^{n_j}(r)$ for which $|a_{n_j}| p^{n_j}(r) = \mu_a(p(r))$.

B. Lepson [5] has introduced the notion of an index $I_o(r)$ of an entire trans-
endental function $f(z)$ and defined

$$I_o(r) = \max_{|a|=r} \nu_a(r|1). \qquad (1.5)$$

172                                    SH. STRELITZ

The index $I_o(r)$ may be bounded only for entire functions of order $\leq 1$ [10].
If $I_o(r) < \infty$ then the function $f(r)$ is called a function of bounded index.

The notion of an index was further generalized in several directions.
In [4] the index

$$I_\alpha(r) = \max_{|a|=r} \nu_a(r|t^\alpha) \qquad (1.7)$$

with a positive constant, in [2] - the index

$$J_\alpha(r) = \max_{|a|\leq r} \nu_a(r|t^\alpha) \qquad (1.8)$$

were introduced. Among the indices $I_\alpha(r)$ and $J_\alpha(r)$ only $J_\alpha(r)$ is always a
non-decreasing function.

The growth of the indices $I_\alpha(r)$, $0\leq\alpha<1$ of entire transcendental solutions
of linear differential equations was investigated in [5] and [10]. In the
present paper we consider analogous problems for entire transcendental solutions of algebraic differential equations. The way we go achieving our
results applies also for the research of analytic functions in an angle. For
sake of brevity and simplicity we, as a rule, consider in this paper entire
solutions of first order algebraic differential equations although similar
results as the ones obtained in this article can be obtained for solutions of
algebraic differential equations of certain classes of higher order too. Our
aim is to show rather how results are obtained than to prove theorems of
general nature, which often can hardly be conveniently formulated, though of
course it can be done.

In order to give an idea of the problems we are dealing with we bring
below the following assertion proven in section 18.

Consider the differential equation

$$\sum_{j=0}^{n-1} P_j(z,\frac{w'}{w}) \; w^{n-j} = g(z) \qquad (1.9)$$

with

$$P_j(z,\eta) = \sum_{i=0}^{m_j} P_{ji}(z) \, \eta^{m_j-i} \; , \quad j=0,1,2,..,n-1,$$

where all the $P_{ji}(z)$ are polynomials. Denote by $\deg P_{ij}$ the degree of the
polynomial $P_{ij}(z)$.

Theorem. Suppose that in (1.9)

$$\deg P_{oo} > \deg P_{om}, \quad m=1,2,\ldots,m_o,$$

$$\deg P_{n-1\ o} > \deg P_{n-1\ m} \quad m=1,2\ldots,\ m_{n-1}$$

(1.10)

and g(z) is an entire function of bounded index.  Then every entire solution
of (1.9) is of bounded index.

It is easy to see by comparison the similarity of this theorem to the one
proven in [10] for linear differential equations.  The similarity will be
even more evident if we remark that the theorem above remains valid if we
change the condition (1.10) by

$$\deg P_{oo} = \deg P_{n-1\ o} \geq \deg P_{ij}$$

for all possible i and j.

Our method of research is based on a certain Wiman-Valrion type theory, we
build up in the present paper for a generalized index (g.i) of an analytic
function in an angle (in particular in the whole complex plane).  Our paragraphs
2,3 and 4 are devoted to the construction of this theory.

## §2.  A Wiman-Valiron Theory for a Generalized Index

2 Definitions.  Let f(z) be a holomorphic function within the angle A':
$|\arg z| < \theta'$.  By A we denote a closed angle $|\arg z| \leq \theta_o < \theta'$ so that $A \subset A'$.

Definition 1 (Basic set).  Let E be a sequence of segments:

$$E = \bigcup_{j=1}^{\infty} \{[R'_j,\ R''_j]\} \ ; \ \{[R'_j,R''_j]\} \cap \{[R'_k,R''_k]\} = \emptyset, \ k \neq j$$

which satisfies the following requirement:

There is a constant h > o such that $R''_j > (1+h)\ R'_j > o, j = 1,2,3\ldots$  .
We call such a sequence of segments a basic set.

Remark that under the conditions of definition 1 the set E is of infinite
logarithmic measure; that is

$$\int_E \frac{dr}{r} = \infty \ .$$

Note also that E may coincide with the whole axis $[o,\infty)$.

Definition 2.  A complex-valued function z=a(r) defined as

$$a(r) = a_j = \text{Const.} \ r \ \varepsilon \ [R'_j,R''_j]\ , \quad j=1,2,3\ldots$$

is called a path on the complex plane on E.

Consider now an analytic function f(z) in A' and a path z = a(r) in the
angle $A \subset A'$ defined on a given basic set E.  The function p(r) introduced in

section 1 must obviously be such that the disc: $|z-a(r)| \leqslant p(r)$, $r\varepsilon E$ belongs to A'. For some reasons, which will be understood later, we restrict the function p(r) a little more requiring:  1)  the existence of a number c > 1 such that $\{|z-a(r)| \leq p(cr)\} \subset \bar{A}'$ for each $r\varepsilon E$ and 2)

$$2)\ \omega(r) = \frac{rp'(r)}{p(r)} \geq \beta_o > o,\ p'(r) = p'(r+o).  \tag{2.1}$$

Under these conditions we define the function

$$I(r|p,a) = \nu_{a(r)}(r|p),\ r\varepsilon E  \tag{2.2}$$

(for the definition of $\nu_a(r|p)$ see section 1).

Definition 3 (Generalized Index).  A function $I(r|p,a)$ defined in (2.2) is called generalized index (g.i) if:

i.  $I(r|p,a)$ is a non-decreasing function on the basic set E,

$$ii.\ \nu_{a_{j+1}}(R'_{j+1}|p) \geq 2\nu_{a_j}(R''|p),\ j=1,2,3..  \tag{2.3}$$

and

$$\lim_{\substack{r\to\infty \\ r\varepsilon E}} I(r|p,a) = \infty  \tag{2.4}$$

Definition 4.  If $I(r|p,a)$ defined in (2.2) is bounded on E: $I(r|p,a) < c < \infty$ then we call $I(r|p,a)$ a bounded generalized index (b g.i).

Examples.  Let f(z) be a transcendental entire function on the bounded complex plane, then
a) the central index $\nu(r)$ of f(z) is a generalized index;
b) the index $I_o(r)$ of (1.6) is not a g.i. because $\omega(r) \equiv 0$ (see (2.1)).  We will use "proximate" to $I_o(r)$ generalized indices $I(r|p,a)$ with certain functions p(r), basic sets E and paths on them a(r), which would represent for some purposes the given $I_o(r)$ good enough;

c) each index

$$J_\alpha(r) = \max_{|a|\leq r} \nu_a(r|t^\alpha),\ \alpha > o  \tag{2.5}$$

constructed for an entire transcendental function in a g.i on the basic set E = { $o,\infty$)} if the intervals $(R'_j, R''_j)$ on which $J_\alpha(r)$ = Const satisfy the condition required in Definition 1.  This condition should not necessarily take place.

The definitions 1-3 are the main concepts needed for the construction
of the Wiman-Valiron Theory for a  g.i  $I(r|p,a)$  in the present paper.  It is
maybe worthwhile to note that a Wiman-Valiron Theory similar to the one we are
building can be obtained by less restricting conditions than the ones here,
which will include also the index $J_\alpha(r)$ from (2.5) without any further requirement.  We will show it at another opportunity.

For our purposes in this paper the requirements imposed on the generalized
index in the three definitions above will suffice.

3.  Notations.  Let us agree upon some additional notations.

a)  The letter C with or without marks on the top will always mean a constant
which may obtain different values at various circumstances although written
identically.

b)  $E_o$ will denote a sequence of intervals on the positive coordinate ray of
bounded logorithmic measure, that is

$$\int_{E_o} \frac{dr}{r} < \infty \ ,$$

with the following property: .every segment $[o,R]$ has common points no more
than with a finite number of intervals belonging to $E_o$.  In various cases $E_o$,
although written identically, may mean different sets.  Evidently a finite
sum of sets $E_o$ is a set $E_o$ too.

c)  E will always denote a basic set.  E with marks on the top as for instance
$E^*$, $\bar{E}$ and so on will mean sequences of segments belonging to E as subsets.
We mark:  $E' = E/E_o$,   $E^{*\prime} = E^*/E_o$,   $\bar{E}' = \bar{E}/E_o$  and so on.

d)  We will often use the notations $I(r|p)$, $I(r)$ or even I instead of the
notation $I(r|p,a)$ if there is no possibility for confusion.

4.  Formulation of the theorems of the Wiman-Valiron theory.  We are now ready
to formulate the main results of this paragraph.

Let $f(z)$ be a holomorphic function in the angle A': $|\arg z| < \theta'$ and let A
be the angle $|\arg z| \le \theta_o < \theta'$: $A \subset A'$.  Suppose further $I(r|p,a) \equiv I(r)$  to be
a g.i. in A with a path $z = a(r)$ within A defined on a basic set E.

Theorem 1.  Let $r\varepsilon E$ be an arbitrary point.

Denote

$$M_{a(r)}(p(r)) = \max_{|z-a(r)|=p(r)} |f(z)| \tag{4.1}$$

and let

$$\omega(r) = \frac{rp'(r)}{p(r)} \ge \beta_o > 0$$

(see (2.1)) (so that under our condition imposed on $p(r)$ in section 2 the disc
$\{|z-a(r)| \leq p(r)\} \subset A')$.

Then on $E' = E/E_o$:

a) $\mu_{a(r)}(p(r) \leq M_{a(r)}(p(r)) \leq \mu_{a(r)}(p(r))\;[I(r)\ln^{1+\alpha}I(r)]^{1/2}$,          (4.2)

where $\alpha > o$ is an arbitrary but fixed constant and $E_o = E_o(\alpha)$.

Remark.  If $\nu_{a(r)}(f/p)$ is bounded on E then on the set $\overset{\infty}{\underset{j=1}{U}}\,[R'_j+\delta,\ R''_j-\delta]$ with
an arbitrary constant $\delta > o$

$$\mu_{a(r)}(p(r) \leq M_{a(r)}(p(r) \leq C_\delta M_{a(r)}(p(r))$$          (4.3)

where $C_\delta$ is a certain constant (depending, obviously, on $\delta$).

Theorem 2.  Under the conditions of Theorem 1 let $\zeta$ be a point on the circle
$|z-a(r)|=p(r)$, $r\varepsilon E$ and let $\xi = \zeta-a(r)$ (so that $|\xi|=|\zeta-a(r)|=p(r)$) be such that

$$|f(\zeta)| = |f(\xi + a(r))| \geq [I(r)]^{-N} M_{a(r)}(p(r));0 \leq N = Const.$$          (4.4)

then (under condition (2.1)) on $E' = E/E_o$

$$f(a(r) + \xi e^\eta) = f(a(r) + \xi)e^{I(r)\eta}(1+\psi_o(r))$$          (4.5)

with

$$|\psi_o(r)| < I^{N+\frac{1}{2}}(r)\ln^{\frac{3+\alpha}{2}}I(r)$$          (4.6)

if only

$$|\eta| \leq [I(r)\ln^{3+\alpha}I(r)]^{-\frac{1}{2}}.$$          (4.7)

Besides for $r\varepsilon E'$

$$\left|(\xi\tfrac{d}{d\xi})^n \ln f(a(r)+\xi)\right| < n!\,I^{nN}(r)\ [I(r)\ln^{3+2\alpha}I(r)]^{\frac{n}{2}}(1+N\ln I(r)),\ n=2,3,4\ldots$$          (4.8)

and

$$\left|\frac{\xi f'(a(r)+\xi)}{f(a(r)+\xi)} - I(r)\right| < 9I^{N+\frac{1}{2}}(r)\ln^{\frac{3+\alpha}{2}}I(r).$$          (4.9)

In particular on $E'$

$$\ln f(a(r)+\xi e^\eta) = \ln f(a(r)+\xi) + I(r)\,\eta + \tilde{\psi}\,\eta^2$$          (4.10)

with

$$|\tilde{\psi}| < [I^{2N+1}(r)\ln^{3+2\alpha}I(r)]\,(1 + N\ln I(r))$$          (4.11)

whenever

$$|\eta| < I^{-N}(r)\,[I(r)\ln^{3+2\alpha}I(r)]^{-\frac{1}{2}}(1+N\ln r)^{-\frac{1}{2}}.$$          (4.12)

As a consequence of theorem 2 we prove

Theorem 3.  Let, under the conditions and notations of Theorem 2, $N<\frac{1}{2}$.  Then

$$\lim_{\substack{r\to\infty \\ r\in E}} \frac{\xi^n f^{(n)}(\zeta)}{f(\zeta)} \cdot \frac{1}{I^n(r)} = 1, \quad n = 1,2,3,\dots \tag{4.13}$$

Theorem 4. Suppose $f(z)$ to be a holomorphic function in the angle $A'$: $|\arg z| < \theta'$. Suppose further all the conditions of Theorem 2 to be satisfied in the closed angle $A \subset A'$: $|\arg z| \leqslant \theta_o < \theta'$. We assume the basic set $E = \overset{\infty}{\underset{j=1}{\cup}} \{[R'_j, R''_j]\}$ to have the following property:

$$R''_j \geqslant R'^{1+\beta}_j, \quad j = 1,2,3\dots \tag{4.14}$$

where $\beta > o$ is an arbitrary fixed constant. Then either

i) there is a subsequence of indices $\{j_k\}$ and a subset

$$\bar{E} = \overset{\infty}{\underset{k=1}{\cup}} \{[\bar{R}'_k, \bar{R}''_k]\} \subset E \text{ such that}$$

$$\{[\bar{R}'_k, \bar{R}''_k]\} \subset \{[R'_{j_k}, R''_{j_k}]\}, \quad k = 1,2,3,\dots \quad \bar{R}'_k = R'^{1+\frac{\beta}{4}}_{j_k},$$

$$\bar{R}''_k = R'^{1+\frac{\beta}{2}}_{j_k} = \bar{R}'^{\frac{1+\frac{\beta}{2}}{1+\frac{\beta}{4}}}_k \quad \text{and on } \bar{E}' = \bar{E}/E_o$$

$$\ln M_{a(r)}(p(r)) > (1+o(1)) \frac{I(r)}{\ln^2 I(r)} \tag{4.15}$$

or

ii) on the subset $\bar{\bar{E}}' = \bar{\bar{E}}/E_o$:

$$\bar{\bar{E}} = \overset{\infty}{\underset{j=1}{\cup}} \{[\bar{\bar{R}}'_j, \bar{\bar{R}}''_j]\}, \quad \bar{\bar{R}}'_j = R'_j, \quad \bar{\bar{R}}''_j = R'^{1+\frac{\beta}{8}}_j = \bar{\bar{R}}'^{1+\frac{\beta}{8}}_j, \quad j = 1,2,3\dots \tag{4.16}$$

there is

$$-(1+o(1)) \frac{I(r)}{\ln^2 I(r)} > \ln M_{a(r)}(p(r)). \tag{4.17}$$

Besides in both cases on the corresponding sets ($\bar{E}'$ or $\bar{\bar{E}}'$)

$$\frac{|\ln M_{a(r)}(p(r))|}{\ln r} \xrightarrow[r\to\infty]{} \infty. \tag{4.18}$$

Remark. We turn the attention of the reader to the fact that a generalized index $I(r|p,a)$ in an angle $A$ may tend to infinity although $f(z)$ is not necessarily unbounded in $A'$. Consider for example the function $e^z$. It is bounded in the angle $A'$: $\frac{3\pi}{4} < \arg z < \frac{5\pi}{4}$. Choose the path $z = a(r)$ to be the negative real axis and $p(r) = \frac{r}{2}$. Obviously, $f(z) = e^{-r} \cdot e^{z+r}$ and $I(r|\frac{t}{2}) = \nu(\frac{r}{2})$, where $\nu(r)$ is the central index of the expansion $e^z = \sum\limits_{n=0}^{\infty} \frac{z^n}{n!}$.

## §3. Auxiliary Lemmas

5. In order to prove the theorem formulated in the previous section 4 we need several lemmas which we state below. The first of them deals with some consequences of the well known Borel-Nevanlinna Theorem concerning the evaluation of the difference $h(x_2) - h(x_1)$ of a non-decreasing function

$$h(x) \xrightarrow[x \to \infty]{} \infty \quad (\text{see }[8]).$$

Lemma 1. Let $h(x) > 0$ be a non-decreasing function on the positive $x$ axis and $\lim\limits_{x \to \infty} h(x) = \infty$. Let further $\alpha, \beta$ and $\gamma$ be arbitrary real numbers with $\alpha > 0$, $\beta$, $\gamma > 0$: $\beta + \gamma \geq 0$. Then for every $x \varepsilon (o,\beta)$ outside may be a certain set $E_o$

$$\left| h(xe^\tau) - h(x) \right| < h^{1-\beta}(x) \ln^{(1-\gamma)(1+\alpha)} h(x) \tag{5.1}$$

if only

$$|\tau| \leq [h^\beta(x) \ln^{\gamma(1+\alpha)} h(x)]^{-1}. \tag{5.2}$$

The proof of this lemma one can find in [11].

6. Lemma 2. Suppose the function $p(r)$ to increase on the segment $[R_1, R_2]$ with a non-decreasing

$$\omega(r) = \frac{rp'(r)}{p(r)} > 0, p'(r) = p'(r+0)$$

(in other words $\ln p(x)$ is a convex function of $\ln x$). Let further $\mu_a(p(r))$ (with a fixed $a$) be the maximal term in the closed disc $|z-a| \leq p(r)$, $R_1 \leq r \leq R_2$ of the expansion (1.3) which we assume converging for $|z-a| \leq p(R_2)$.

Then

$$\frac{p(r)\mu_a{}'(p(r))}{\mu_a(p(r))} \equiv \mu_a(r|p), \quad R_1 \leq r < R_2 \tag{6.1}.$$

and

$$\omega(r)\, \nu_a(r\,|\,p)\tau \;\leq\; \ln\mu_a(p(r)e^{\omega(r)\tau}) \;-\; \ln\mu_a(P(r)) \;\leq\; \omega(r)\nu_a(re^{\tau}\,|\,p)\tau \tag{6.2}$$

for each $\tau$ such that $p(R_1) \leq p(r)e^{\omega(r)\tau} \leq p(R_2)$.

Proof.  Consider the expansion

$$f(z) = \sum_{n=0}^{\infty} A_n (z-a)^n \tag{6.3}$$

which we assume converging in the disc $|z-a| \leq p(R_2)$.

Then

$$\mu_a(p(r)) = |A_\nu| p^{\nu} (r)$$

where $\mu_a(R)$ is the maximal term of (6.3) on the circle $|z-a|=R$.

But

$$\frac{R\mu_a'(R)}{\mu_a(R)} = \nu_a(R) \tag{6.4}$$

(see for example [12]).  Substitute $p(r)$ for R we obtain (6.1).
In view of (6.4)

$$\ln\mu_a(r'') - \ln\mu_a(r') = \int_{r'}^{r''} \frac{\nu_a(t)}{t}\, dt,$$

whence

$$\nu_a(r')\, \ln\frac{r''}{r'} \;\leq\; \ln\mu_a(r'') - \ln\mu_a(r') \;\leq\; \nu_a(r'')\, \ln\frac{r''}{r'}. \tag{6.5}$$

Putting here $r'= p(r)$ and $r'' = p(r)e^{\omega(r)\tau}$ we obtain

$$\nu_a(p(r))\omega(r)\tau \;\leq\; \ln\mu_a(p(r)e^{\omega\tau}) - \ln\mu_a(p(r)) \;\leq\; \omega(r)\nu_a(p(r)e^{\omega(r)\tau})\tau. \tag{6.6}$$

But $\ln p(r)$ is a convex function of $\ln r$ and therefore

$$\omega(r)\tau \;\leq\; \ln p(re^{\tau}) - \ln p(r) \;\leq\; \omega(re^{\tau})\tau. \tag{6.7}$$

Now remember that $\nu_a(p(R)) = \nu_a(R\,|\,p)$ is a non-decreasing function.

Therefore in view of (6.7) $p(r)e^{\omega(r)\tau} \leq \ln p(re^{\tau})$ and (6.2) follows from (6.6) with respect to (6.7).

7.  Lemma 3.  Suppose all the conditions of Lemma 2 to be satisfied.  Let

$$M_a(p(r)) = \max_{|z-a|=p(r)} |f(z)| \quad \text{and} \quad K_a(R) = \frac{RM_a'(R'+0)}{M_a(R)}. \tag{7.1}$$

Then

$$\omega(r)\, K_a(p(r))\tau \le \ln M_a(p(r)e^{\omega(r)\tau}) - \ln M_a(p(r)) \le \omega(r)\, K_a(p(re^\tau))\tau. \qquad (7.2)$$

Proof.  Note that accoridng to Hadamard's three circles Theorem

$$\frac{d\ln M_a(R)}{d\ln R} = K_a(R) \text{ is a non-decreasing function.}$$  On ground of this fact

the proof of the lemma is literally the same as that of Lemma 2:  it is

enough to replace there $\nu_a(r|p)$ by $K_a(p(r))$ and $\mu_a(p(r))$ by $M_a(p(r))$.

4.  Proofs of the Theorems

8.  Proof of Theorem 1.  Let

$$f(z) = \sum_{n=0}^{\infty} A_n(z-a(r))^n; \quad A_n = \frac{f^{(n)}(a(r))}{n!}, \quad n=0,1,2\ldots \qquad . \qquad (8.1)$$

This series converges in view of the properties of the function $p(r)$
indicated in section 2 in the disc $|z-a(r)| < p(cr)$ with a certain $c > 1$.
The left inequality of (4.2) is a consequence of the Cauchy inequalities
for the Taylor coefficients $A_n$.  Thus it remains to prove the right

inequalities in (4.2).  According to (2.2)

$$I(r|p,a) \equiv I(r) = \nu_{a(r)}(r|p), \quad r\varepsilon E$$

where $|z-a(r)| = p(r)$.  From (8.1) we conclude that

$$M_{a(r)}(p(r)) \le \sum_{n=0}^{\infty} |A_n| p^n(r). \qquad (8.2)$$

We rewrite the series (8.2) as a sum of two parts:

$$M_{a(r)}(p(r)) \le \sum_{n=0}^{\nu_{a(r)}(r|p)} |A_n| p^n(r) + \sum_{n=\nu_{a(r)}(r|p)+1}^{\infty} |A_n| p^n(r) = \sigma_1 + \sigma_2. \qquad (8.3)$$

Our next aim is to evaluate the sums $\sigma_1$ and $\sigma_2$.  For simplicity denote

$\nu_{a(r)}(r|p) = \nu_a$.  Let $\tau > 0$ be an arbirtary small enough number.  We have:

$$\sigma_2 = \sum_{n=\nu_a+1}^{\infty} |A_n| p^n(r) = \sum_{q=1}^{\infty} |A_{\nu_a+q}| p^{\nu_a+q}(r) =$$

$$= \sum_{q=1}^{\infty} |A_{\nu_a+q}| p^{\nu_a+q}(r) e^{(\nu_a+q)\omega(r)\tau} e^{-(\nu_a+q)\omega(r)\tau}, \quad \omega(r)=rp'(r)/p(r). \tag{8.4}$$

But according to (6.7)

$$p(r)e^{\omega(r)\tau} < p(re^{\tau}), \tag{8.5}$$

so that from (8.4) it follows

$$\sigma_2 \leq \sum_{q=1}^{\infty} |A_{\nu_a+q}| p^{\nu_a+q}(re^{\tau}) e^{-(\nu_a+q)\omega(r)\tau} \leq \mu_a (p(re^{\tau})) e^{-\nu_a\omega(r)\tau} \times$$

$$\sum_{q=1}^{\infty} e^{-q\omega(r)\tau}$$

where we put $a(r) = a$.

Further, according to (6.6)

$$\sigma_2 \leq \mu_a (p(re^{\tau})) e^{-\nu_a\omega(r)\tau} [1 - e^{\omega(r)\tau}]^{-1}$$

$$\leq \mu_a (p(r)) e^{\omega(r)[\nu_a(re^{\tau}|p) - \nu_a(r|p)]\tau} [1-e^{-\omega(r)\tau}]^{-1} =$$

$$= \mu_a (p(r) e^{[\omega(r) I(re^{\tau})-I(r)]\tau} [1-e^{-\omega(r)\tau}]^{-1} \tag{8.6}$$

The function $I(r)$ is defined on the basic set

$$E = \bigcup_{u=1}^{\infty} \{ [R_j', R_j''] \}, \quad R_j'' > (1+h) R_j', \quad h > 0 \text{ and}$$

$$\{ [R_j',R_j''] \} \cap \{ [R_k',R_k''] \} = \emptyset \quad \text{if } j \neq k. \quad \text{For } r\epsilon (R_j'', R_{j+1}')$$

We define additionally

$$I(r) = \frac{I(R_j'')+I(R_{j+1}')}{2}, \quad j = 1,2,3\ldots \quad .$$

Thus we obtain a non-decreasing function on the whole positive axis.
According to Lemma 1 with $\beta=\gamma=\frac{1}{2}$ outside a certain set $E_o$

$$|I(re^{\tau}) - I(r)| < (I(r)\ln^{1+\alpha} I(r))^{\frac{1}{2}} \tag{8.7}$$

whenever

$$|\tau| \leq \omega(r)^{-1} |I(r)\ln^{1+\alpha} I(r)]^{-\frac{1}{2}} \tag{8.8}$$

with an arbitrary but fixed $\alpha > 0$ and $E_o = E_o(\alpha)$.

182     SH. STRELITZ

It is important to note, that because of (2.1) for each j there are neighborhoods of $R_j'$ and $R_j''$ belonging to $E/E_o$ such that if $r\varepsilon E' = E/E_o$ then also $\{[re^{-\tau}, re^{\tau}]\} \subset E$; with the values $\tau$ of (8.8). In view of (8.6) since

$$\omega(r) \geq \beta_o > 0$$

$$\sigma_2 \leq \mu_{a(r)}(p(r))(1 - e^{\omega(r)\tau})^{-1} < \tfrac{1}{2}\mu_{a(r)}(p(r))[I(r)\ln^{1+\alpha}I(r)]^{\frac{1}{2}}. \tag{8.9}$$

Now we start with the evaluation of $\sigma_1$ from (8.5). On ground of (8.5) we get:

$$\sigma_1 = \sum_{n=0}^{\nu_a}|A_n|p^n(r) = \sum_{n=0}^{\nu_a}|A_n|p^n(r)\, e^{-n\omega(r)\tau}\, e^{n\omega(r)\tau} \leq$$

$$\leq \sum_{n=0}^{\nu_a}|A_n|p^n(re^{-\tau})\, e^{n\omega(r)\tau} \leq \mu_{a(r)}(p(re^{-\tau}))\, e^{\nu_a\omega(r)\tau}(1-e^{-\omega(r)\tau})^{-1} \leq$$

$$< \mu_{a(r)}(p(r))\, e^{\omega(r)[\nu_a(r|p)-\nu_a(re^{-\tau}|p)]\tau} \cdot (1-e^{-\omega(r)\tau})^{-1} =$$

$$= \mu_{a(r)}(p(r))e^{\omega(r)[I(r)-I(re^{-\tau})]\tau}(1-e^{-\omega(r)})^{-1}. \tag{8.10}$$

In the same manner as we proved (8.9) now we obtain also by Lemma 1

$$\sigma_1 < \tfrac{1}{2}\mu_{a(r)}(p(r))[I(r)\ln^{1+\alpha}I(r)]^{\frac{1}{2}}. \tag{8.11}$$

From (8.9) and (8.11) we get (4.2).

Proof of the remark to Theorem 1. Consider now the bounded function $\nu_{a(r)}(r|p) < \infty$ on E. Then (4.3) follows immediately from (8.6) and (8.10).

9. Lemma 4. Assume that all the conditions of Theorem 1 are satisfied. Let $K_a(r)$ be the function defined in (7.1). Then on $E' = E/E_o$

$$K_{a(r)}(p(r)) = I(r) + \Psi(I(r)\ln^{3+\alpha}I(r)) \tag{9.1}$$

where

$$|\Psi(I(x))| < \sqrt{x}. \tag{9.2}$$

Proof. According to (7.2) and (4.2) we have on E':

$$\omega(r)K_{a(r)}(p(r)\tau \leq \ln M_{a(r)}(p(r)e^{\omega(r)\tau}) - \ln M_{a(r)}(p(r)) <$$

$$< \ln \mu_{a(r)}(p(r)e^{\omega(r)\tau}) - \ln \mu_{a(r)}(p(r)) + \tfrac{1}{2}\ln[I(re^{\tau})\ln^{1+\alpha}I(re^{\tau})]. \tag{9.3}$$

By (6.2)

$$\omega(r)K_{a(r)}(p(r))\tau < \omega(r)I(re^{\tau}) + \tfrac{1}{2}\ln\,[I(re^{\tau})\ln^{1+\alpha}I(re^{\tau})]\,. \qquad (9.4)$$

Take

$$\omega(r)\tau \;=\; [I(r)\ln^{1+\frac{\alpha}{2}}I(r)]^{-\frac{1}{2}}. \qquad (9.5)$$

According to Lemma 1 with $\beta = \gamma = \tfrac{1}{2}$ on $E'$ since $\omega(r) \geq \beta_o > 0$

$$I(re^{\tau}) < I(r) + [\,I(r)\ln^{1+\alpha}]^{\frac{1}{2}}$$

and subsequently with respect to (9.4)

$$K_{a(r)}(p(r)) < I(r) + [I(r)\ln^{3+\alpha}I(r)]^{\frac{1}{2}}, \quad r\varepsilon E'. \qquad (9.6)$$

On the other hand from (7.2) and (4.2) it follows

$$\omega(r)K_{a(r)}(p(r))\tau \geq \ln M_{a(r)}(p(r)) - \ln M_{a(r)}(p(r)e^{-\omega(r)\tau}) \geq$$

$$\geq \ln\mu_{a(r)}(p(r)) - \ln\mu_{a(r)}(p(r)e^{-\omega(r)\tau}) - \tfrac{1}{2}\ln[I(re^{-\tau})\ln^{1+\alpha}I(re^{-\tau})] >$$

$$> \omega(r)I(r)\tau - \tfrac{1}{2}\ln[I(re^{\tau})\ln^{1+\alpha}I(re^{-\tau}] \geq$$

$$\geq \omega(r)I(r)\tau - \tfrac{1}{2}\,\ln[I(r)\ln^{1+\alpha}I(r)]$$

since $I(r)$ is a non-decreasing function.

If the product $\omega(r)\tau$ is the one from (9.5) then

$$K_{a(r)}(p(r)) > I(r) - [I(r)\ln^{3+\alpha}I(r)], \quad r\varepsilon E'. \qquad (9.7)$$

The relations (9.6) and (9.7) prove the lemma.

10. Lemma 5.  Assume that all the conditions of Theorem 1 are satisfied.  Let $r\varepsilon E$ and let $\zeta$ be a point on the circle $|a(r)-z| = p(r)$, where

$$|f(\zeta)| \geq I^{-N}(r)\,M_{a(r)}(p(r)) \qquad (10.1)$$

with a fixed $N > 0$.  Then for $|\eta| \leq \tau\ln^{-1}I(r)$
with

$$\tau = [I(r)\ln^{1+\alpha}I(r)]^{-\frac{1}{2}}$$

on $E' = E/E_o$ denoting $\zeta = a(r) + \xi$

$$\frac{|f(a(r)+\xi e^{\eta})|}{|f(a(r)+\xi)|}\; e^{-K_{a(r)}(p(r))} < cI^N(r) \qquad (10.2)$$

where $c > 0$ is a constant ($c < 3$) independent upon $r$.

Proof.  Denote $\eta = \omega(r)\,(\tilde{\tau}+i\tau)$.  We have

$$\ln\frac{|f(a(r)+\xi e^{\eta})|}{|f(a(r)+\xi)|} \leq \ln M_{a(r)}(p(r)e^{\omega(r)\tilde{\tau}}) - \ln M_{a(r)}(p(r)) + N\ln I(r).$$

184                        SH. STRELITZ

According to (7.2)

$$\ln\left|\frac{f(a(r) + \zeta e^{\eta})}{f(a(r) + \xi)}\right| - K_{a(r)}(p(r))\tilde{\tau}\omega(r) \leq$$

$$\leq |K_{a(r)}(p(re^{\tilde{\tau}})) = K_{a(r)}(p(r))| \; |\tilde{\tau}|\omega(r) + N\ln I(r).$$

For $r\epsilon E'$ with respect to Lemma 4

$$\frac{\ln|f(a(r) + \xi e^{\eta})|}{|f(a(r) + \xi)|} - K_{a(r)}(p(r))\tilde{\tau}\omega(r) \leq$$

$$\leq |I(re^{\tilde{\tau}}) - I(r)| \; |\tilde{\tau}|\omega(r) + |\tilde{\tau}|\psi + N\ln I(r) \qquad (10.3)$$

where

$$\psi = 0((I(r)\ln^{3+\alpha} I(r))^{1/2}). \qquad (10.4)$$

If

$$\omega(r)|\tilde{\tau}| \leq [I(r)\ln^{1+2\alpha} I(r)]^{-\frac{1}{2}}$$

then (10.2) follows from (10.3) in view of Lemma 1.

Remark. From the proof of the last lemma it follows that

$$\frac{M_{a(r)}(p(r)e^{\omega(r)\tilde{\tau}})}{M_{a(r)}(p(r))} \; e^{-K_{a(r)}(p(r))\tilde{\tau}} < C \qquad (10.5)$$

with $\tilde{\tau}$ from (10.4).

11. Proof of Theorem 2. Consider the function

$$F(\eta) = \frac{f(a(r) + \xi e^{\eta})}{f(a(r) + \xi)} \; e^{-I(r)\eta} = 1 + \sum_{j=1}^{\infty} B_j \eta^j; \quad I(r) \equiv I(r|p,a) \qquad (11.1)$$

where $r\epsilon E'$. As indicated in Section 2 the series (11.1) converges in the disc

$$|\eta| \leq [I(r)\ln^{3+\alpha} I(r)]^{-\frac{1}{2}} \qquad (11.2)$$

for large enough r (indeed the function f(z) is analytic in the disc $|\zeta - a(r)| \leq p(cr)$ where c > 1 is a constant not depending on r, all the more so in the

disc $|\tau - a(r)| \leq p(r) \; e^{|\eta|\omega(r)}$ since $p(r)e^{|\eta|\omega(r)} < p(re^{|\eta|}) < p(cr)$ while

$\eta$ satisfies; (11.2) see (6.7)).

From (10.2) because of (9.1) with (9.2) it follows

$$\left| \frac{f(a(r) + \xi e^{\eta})}{f(a(r) + \xi)} \; e^{-I(r)\eta} \right| < 9I^{N}(r) \qquad (11.3)$$

for $\eta$ from (11.2).  By Schwarz's Lemma

$$\left| \sum_{j=1}^{\infty} B_j \eta^j \right| < (9I^{N}(r) + 1) \; (I(r)\ln^{3+\alpha} I(r))^{\frac{1}{2}} |\eta|. \qquad (11.4)$$

Thus in the disc (11.2), if only $r \varepsilon E'$,

$$F(\eta) = 1 + \psi_o \eta$$

with $\psi_o$ satisfying the inequality

$$|\psi_o| < I^{N+\frac{1}{2}}(r)\ln^{\frac{3+\alpha}{2}} I(r).$$

Hence (4.6) is correct and

$$|F(\eta)| < I^{-N-\frac{1}{2}}(r) \; \ln^{-3-\alpha} I(r) \qquad (11.5)$$

provided

$$|\eta| < I^{-N-\frac{1}{2}}(r) \; \ln^{-3-\alpha} I(r). \qquad (11.6)$$

that is the function $f(a(r) + \xi e^{\eta})$ does not vanish in the disc (11.6).
Besides, from (11.4) and (11.2) it follows by the Cauchy evaluation of
the Taylor coefficients that

$$|B_1| = \left| \frac{\xi f'(a(r) + \xi)}{f(a(r) + \xi)} - I(r) \right| < 9I^{N+\frac{1}{2}}(r) \; \ln^{-\frac{3+\alpha}{2}} I(r)$$

which coincides with the relation (4.9).

Within the disc (11.5) the function $F_1(\eta) = \ln F(\eta)$ (more precisely: each
branch of it) is a holomorphic one (for certainty we choose the branch of
the logarithm with $\ln F(0) = 0$).  Therefore there is a power expansion

$$F_1(\eta) = \ln f(a(r) + \xi e^{\eta}) - \ln f(a(r) + \xi) - I(r)\eta = \sum_{j=1}^{\infty} c_j \cdot \eta^j \qquad (11.7)$$

converging for $\eta$ of (11.6), where evidently

$$c_j = \frac{1}{j!} \; (\xi \frac{d}{d\xi})^j \ln f(a(r) + \xi), \qquad j = 1,2,3,\ldots \; . \qquad (11.8)$$

We need now the following evaluation of the Taylor coefficients depending on
the upper bound of the real component of the given function ([3], p. 199):

let $g(z) = \sum_{n=0}^{\infty} a_n z^n$

be a Taylor series converging for $|z| < R$.  If $\mathrm{Reg}(z) \leq u$ then

$$|a_n| \leq \frac{2(u - \mathrm{Rea}_o)}{R^n} \, , \quad n = 1,2,3,\dots \quad .$$

In our case $\mathrm{ReF}_1(\eta) < 9 + N \ln I(r)$ whenever $\eta$ satisfies (11.6) and therefore

putting $\zeta = a(r) + \xi$ in view of (11.3)

$$j! \, |c_j| = |\xi(\tfrac{d}{d\xi})^j \ln f(\zeta)| < j! \, [I^{N+\frac{1}{2}}(r) \ln^{\frac{3+\frac{3}{2}\alpha}{2}} I(r)]^j (1+N\ln I(r)), \qquad (11.9)$$

$$j = 2,3,4,\dots \quad .$$

On the other hand

$$\ln f(a(r) + \xi e^\eta) - \ln f(a(r)+ + \xi) - K_{a(r)}(p(r))\eta = \eta^2 \sum_{j=2}^{\infty} c_j \eta^{j-2} =$$

$$= \eta^2 \psi^*(\eta).$$

Let

$$|\eta| \leq [I^{N+\frac{1}{2}}(r)\ln^{3+\alpha'} I(r)]^{-\frac{1}{2}} (1+N\ln I(r))^{-\frac{1}{2}}$$

with $\alpha' > \frac{3}{2}\alpha$.  Then according to (11.9)

$$|\psi^*(\eta)| \leq \sum_{j=2}^{\infty} |c_j| \, |\eta|^{j-2} \leq (I^{2N+1}(r)\ln^{3+\alpha'} I(r)) \, (1+N\ln I(r) \sum_{n=0}^{\infty} \frac{1}{\ln^{(\alpha'-\alpha)} I(r)^n} .$$

Replacing here $\alpha'$ by $2\alpha$ we obtain (4.12) and thereby the completion of the

proof of Theorem 2.

12.  Proof of Theorem 3.  Evidently

$$\frac{z^2 f''(z)}{f(z)} \equiv (\frac{zf'}{f})^2 - \frac{zf'}{f} + z(\frac{zf'}{f})' . \qquad (12.1)$$

It is not difficult to verify by induction, that

$$\frac{z^m f^{(m)}(z)}{f(z)} \equiv (\frac{zf'}{f})^m + \sum_{\substack{p i_p = m, \, i_p < m}} B_{i_1 \dots i_p} \prod_{P}[(z\frac{d}{dz})^P \ln f(z)]^{i_p} \qquad (12.2)$$

with certain constants $B_{i_1 i_2 \dots i_p}$ .  Let $f(z)$ be an analytic function in

the angle A' and let further $\zeta$ be a point on the circle $|z-a(r)|=p(r)$

with $\xi=\zeta-a(r)$ such that for $r\varepsilon E'$ (4.4) is satisfied and thereby (4.8) is

correct.  We have:

$$\Pi^* = \Pi_{\Sigma p i_p = m, i_p < m} \left| (\xi \tfrac{d}{d\xi})^P \ln f(a(r)+\xi) \right|^{i_P} = \left| \frac{\xi f'(\zeta)}{f(\zeta)} \right|^{i_I} \Pi_{\substack{\Sigma p i_p = m i_1, i_p < m \\ p > 1}} \left| (\xi \tfrac{d}{d\xi})^P \ln f(a(r)+\xi) \right|^{i_P} . \qquad (12.3)$$

Using (4.8) and (4.9) we now obtain:

$$\Pi^* = \Pi_{\Sigma p i_p = m, i_p < m} \left| (\xi \tfrac{d}{d\xi})^P \ln f(a(r)+\xi) \right|^{i_P} < \left[ I(r) + 9 I^{N+\frac12}(r) \ln^{\frac{3+\alpha}{2}}_{\substack{\Sigma p i_p \\ p > 1}} I(r) \right]^{i_j}$$

$$\cdot \; \Pi_{\Sigma p i_p = m - i_1, i_p < m} (2!^{i_2} 3!^{i_3} .. p!^{i_P}) (1 + n \ln I(r))^{i_2 + i_3 + i_P} [I^{N+1/2}(r) \ln I(r)]$$

Put now $N < \frac12$.  Then

$$\Pi^* < C_{i_1 i_2 \ldots i_p} [I(r)]^{i_1 + (\beta + \frac12) \Sigma p i_p \atop p > 1} \qquad (12.4)$$

where $C_{i_1 i_2 \ldots i_p}$ is a certain constant, $N < \beta < \frac12$ and $r \in E'/E_o, E_o = E_o(\beta)$.  But

$$i_1 + (\beta + \tfrac12) \sum_{p > 1} p i_p = (\tfrac12 - \beta) i_1 + (\tfrac12 + \beta)(i_1 + \sum_{m \geq p > 1} p i_p) = (\tfrac12 - \beta) i_1 + (\tfrac12 + \beta) m$$

since $\Sigma i_p = m$.  Consequently according to (12.4)

$$\Pi^* < C_{i_1 i_2 \ldots i_p} I(r)^{(\frac12 - \beta) i_1 + (\frac12 + \beta) m} \qquad (12.5)$$

Note that $i_1 < m-1$, because assuming $i_1 = m-1$ the equation $\sum_{p > 1} p i_p = 1$ will have no positive integers as solutions.  So $i_1 \leq m-2$ and in view of (12.5)

$$\Pi^* \leq C_{i_1 i_2 \ldots i_p} [I(r)]^{m-1+\beta}; \quad \beta < 1/2. \qquad (12.6)$$

From (12.2) it thus follows according to (12.6)

$$\left| \frac{\xi^m f^{(m)}(\zeta)}{f(\zeta)} - \left( \frac{\xi f'(\zeta)}{f(\zeta)} \right)^m \right| < \sum_{\Sigma p i_p = m, i_p < m} B_{i_1 i_2 \ldots i_m} \left| C_{i_1 i_2 \ldots i_p} \right| [I(r)]^{m-1+\beta} =$$

$$= C_m [I(r)]^{m-1+\beta}; \quad C_m = \sum_{\Sigma p i_p = m, i_p < m} \left| B_{i_1 i_2 \ldots i_m} \right| \left| C_{i_1 i_2 \ldots i_m} \right|; \quad \beta < \tfrac12 .$$

Now by (4.9)

$$\left| \frac{\xi^m f^{(m)}(\zeta)}{f(\zeta)} - I^m(r) \right| < \left| \left( \frac{\xi f'(\zeta)}{f(\zeta)} \right)^m \right| - I^m(r) \;\; +$$

$$+ C_m I(r)^{m-1+\beta} < 9I(r) + C_m I(r)^{m-1+\beta}.$$

Since $\beta < \tfrac{1}{2}$ the proof of Theorem 3 follows from the last inequality $r$ tending to $\infty$ on E.

13.  Proof of Theorem 4.  a) Suppose

$$\ln M_{a(R_1^!)}(R_1^! {}^{1+\frac{\beta}{4}})) \geq 1 \tag{13.1}$$

Let $N_o = \{j_k\}$ be a subsequence of the sequence N of the natural numbers such that being $i = j_k$ (13.1) is correct.  Then on every segment

$$[R_i^! {}^{1+\frac{\beta}{4}}, R_i^"] \text{ with } i \varepsilon N_o \quad \ln M_{a(R_i^!)}(p(r)) \quad 0.$$ According to (7.2), (9.1) and

(9.2) on the set of points $\{[R_i^! {}^{1+\frac{\beta}{4}}, R_i^"]\}/E_o$

$$(1+o(1))I(re^{-\tau})\tau \leq \ln M_{a(R_i^!)}(p(r)) - \ln_{a(R_i^!)} M\{p(r)e^{-\omega(r)\tau}) <$$

$$< \ln M_{a(R_i^!)}(p(r)). \tag{13.2}$$

In view of Lemma 1 taking there $\beta = 0$, (remark:  not the $\beta$ in (13.1))

$$\gamma(1+\alpha) = 2 \text{ and } \tau = \frac{1}{\omega(r)\ln^2 I(r)}$$

$$(1+o(1)) \frac{I(r)}{\ln^2 I(r)} \leq \ln M_{a(R_i^!)}(p(r)), \quad r\varepsilon\{[R_j^! {}^{1+\frac{\beta}{4}}, R_j^"]\}/E_o.$$

Thus we have (4.15).

b)  Let (13.1) be wrong for all $i \geq 0$.

Then $\ln M_{a(R_j^!)}(p(r)) < 0$ whenever $r\varepsilon [R_j^!, R_j^! {}^{1+\frac{\beta}{4}}]$.  Now with respect to (13.2)

$$(1+o(1))\omega(r)I(r)\tau \leq \ln M_{a(R_j^!)}(p(r)e^{\omega(r)\tau}) - \ln M_{a(R_n')}(p(r)) <$$

$$< - \ln M_{a(R_j^!)}(p(r)).$$

Taking

$$\tau = \frac{1}{\omega(r)\ln^2 I(r)}$$

We get (4.17).

Suppose now the case a) to take place.  Consider the function $I(r)$ on the set $E = \overset{\infty}{\underset{k=1}{\cup}} [R'_{j_k}{}^{1+\frac{\beta}{4}}, R'_{j_k}{}^{1+\frac{\beta}{2}}]$.  For $r \in E$ $[R'_p{}^{1+\frac{\beta}{4}}, R'_p{}^{1+\frac{\beta}{2}}]$ we have

$r^{\frac{1+\beta}{1+\frac{\beta}{2}}} \in [R'_p{}^{1+\frac{\beta}{2}}, R''_p]$.  Since $R''_p \geq R'_p{}^{1+\beta}$ (7.2) shows, that

$$(1+o(1))\omega(r)I(r)\,\frac{\beta}{2+\beta}\,\ln r \leq$$

$$\leq \ln M_{a(R'_p)}(p(r)e^{\omega(r)\frac{2+2\beta}{2+\beta}\ln r}) - \ln M_{a(R'_p)}(p(r))$$

$$< \ln M_{a(R'_p)}(p(r))e^{\omega(r)\frac{2+2\beta}{2+\beta}\ln r}) \leq \ln M_{a(R'_j)}(p(r^{1+\frac{2+2\beta}{2+\beta}})),$$

so that (4.5) is correct on $\bar{E}'$.

Let us now consider the case b) and the set

$$\bar{E} = \underset{j \geq j_o}{\cup} \{[R'_j, R'_j{}^{1+\frac{\beta}{4}}]\}.$$  If $r \in [R'_j, R'_j{}^{1+\frac{\beta}{8}}] \subset \bar{E}$, then for $j$ large enough

$$r^{\frac{1+\frac{\beta}{8}}{1+\frac{\beta}{4}}} \in [R'_j{}^{1+\frac{\beta}{8}}, R'_j{}^{1+\frac{\beta}{4}}]$$

and in view of (13.2)

$$(1+o(1))\,\frac{8+2\beta}{8+\beta}\,\omega(r)I(r)\ln r \leq$$

$$\leq \ln M_{a(R'_j)}(p(r)e^{\omega(r)\frac{8+2\beta}{8+\beta}\ln r}) - \ln M_{a(R'_j)}(p(r)) \leq$$

$$\leq -\ln M_{a(R'_j)}(p(r))$$

if only $r \in \bar{E}'$.  Thus (4.17) holds in this case too.

§5.   On Entire Solutions of Bounded Index of First Order Algebraic Differential

Equations.

14.  In this paragraph we consider the index

$$I_o(r) = \max_{|a|=r} \nu_a^{\cdot}(r|1).$$                                    (14.)

An entire function, as we know (see §1), is of bounded index if

$I_o(r) \le N < \infty$.  Otherwise N can be defined equivalently as the maximal number

such that for every complex z:   $|z| < \infty$

$$\max_{1 \le k \le N}\{|\frac{f^{(k)}(z)}{k!}|\} \ge |\frac{f^{(m)}(z)}{m!}| \; ; \; m = 0,1,2,3... \; .$$     (14.2)

In [10] , [1] was considered a linear differential equation

$$P_o(z)w^{(n)} + P_1(z)w^{(n-1)} +...+P_n(z)w = g(z)$$                          (14.3)

with all the $P_j(z)$ as polynomials and with an entire function g(z) of

bounded index.  There was posed the problem to find conditions under which all

the entire transcendental solutions of (14.3) are of bounded index.  In this

paragraph a similar problem for first order alegebrai differential equation is

considered.  We would like to note again that our method is apt also for

algebraic differential equations of higher finite orders for which the usual

Wiman-Valiron method applies for the estimation of the growth of entire

transcendental solutions; in particular our method can be successfully used

for the investigation of the problem mentioned above in case of equation (14.3).

We will return to this equation in §7.

15.  Consider the differential equation

$$F_o(z,w,w') \equiv \sum_{i+j=0}^{n} Q_{ij}(z)w^i w'^j = 0,$$

where all the $Q_{ij}(z)$ are polynomicals.  For conveniency we rewrite this

equation in the form

$$\sum_{j=0}^{n} P_j(z,\frac{w'}{w})w^{n-j} = 0$$                               (15.1)

$$\text{with } P_j(z,\eta) = \sum_{i=0}^{m_j} P_{ji}(z)\eta^{m_j-i}$$

and $P_{ji}(z) = A_{ji}(1+o)(1))z^{\alpha_{ji}}.$ 

$$(15.2)$$

Denote by $\deg P_{ji}$ the degree of the polynomial $P_{ji}(z)$. The following assertion holds.

Theorem 5. Suppose that in (15.1)

$$\deg P_{oo} > \deg P_{nm}, \quad m=1,2\ldots,m_o$$

$$(15.3)$$

and

$$\deg P_{no} > \deg P_{nm}, \quad m=1,2\ldots, m_n.$$

$$(15.4)$$

Then each entire transcendtal solution of (15.1) is of bounded index.

Proof. Let the solution $w(z)$ of (14.1) be an entire transcendental function of unbounded index in contrary with the assertion of the Theorem, that is $\overline{\lim_{r\to\infty}} I_o(r) = \infty$ .

Let $\{r_j\}$, $r_j \uparrow \infty$ be a sequence such that

$$\overline{\lim_{r\to\infty}} I_o(r) = \lim_{j\to\infty} I_o(r_j) = \infty.$$

$$(15.5)$$

Choose a number $\gamma > 0$ with the properties

$$\deg P_{oo} - m_o \gamma < \deg P_{oj} - j\gamma , \quad j=1,2,\ldots,m_o$$

$$\deg P_{no} - m_n\gamma > \deg P_{nj} - j\gamma, \quad j=1,1,\ldots,m_n$$

Note that $\gamma$ can be taken arbitrarily small (but fixed). There exists a sequence $\{a_j\}$ of points on the complex plane such that $|a_j| = R_j$; $j=1,2,3\ldots$

and

$$I_o(R_j) = \max_{|a|=R_j} \nu_a(R_j|1) = \nu_{a_j}(R_j|1), \quad j = 1,2,3\ldots .$$

$$(15.6)$$

Assume $E = \bigcup_{j=1}^{\infty} \{[R_j, R_j^{1+\beta}]\}$ with a fixed but arbitrary $\beta>0$ to be a basic set and the sequence $R_j$ to be so rare that

$$\{[R_j,R_j^{1+\beta}]\} \cap \{[R_k,R_k^{1+\beta}]\} = \emptyset$$

and

$$2\nu_{a_j}(R_j^{1+\beta}|t^\gamma) < \nu_{a_{j+1}}(R_{j+1}|t^\gamma)$$

(for the definition of $\nu_a(r|p)$ see Section 1).

192                           SH. STRELITZ

Under these assumptions we define on E the path

$$a(r) = a_j; \quad R_j \leq r \leq R,^{1+\beta}, \quad j=1,2,3,\ldots$$

and then the g.i

$$I(r|t^\gamma,a) \equiv I(r) \equiv \nu_{a(r)}(r|t^\gamma), \quad r\varepsilon E.$$

According to Theorem 4 there is either a subset $\bar{E}$ or a subset $\bar{\bar{E}}$ such that either on $\bar{E}'$ (4.15) or on $\bar{\bar{E}}'$ (4.17) is satisfied.

Let $\zeta$ be a point on the circle

$$|z-a(r)| = \tau^\gamma \text{ where } |w(\zeta)| = M_{a_j}(\tau^\gamma). \quad \text{By (4.13) either on } \bar{E}' \text{ or on } \bar{\bar{E}}''$$

$$w'(\zeta) - (1+o(1))\frac{I}{\xi} w(\zeta)$$

with $I = I(r)$, $|\zeta - a(r)| = \tau^\gamma$ and $\xi = \zeta - a(r)$, so that $|\xi| = r^\gamma$.

Replacing z by $\zeta$ in (15.1) and using the last equality we obtain:

$$\sum_{j=1}^{n} \sum_{i=0}^{m_j} P_{ji}(\zeta) \left(\frac{I}{\xi}\right)^{m_j-i} (1+o(1))w^{n-j}(\zeta) = 0. \tag{15.7}$$

Theorem 4 shows that either

$$\sum_{i=0}^{m_n} P_{ni}(\zeta) \left(\frac{I}{\xi}\right)^{m_n-i} (1+o(1)) = o(1); \quad |\zeta| = r\varepsilon\bar{\bar{E}}' \tag{15.8}$$

or

$$\sum_{i=0}^{m_o} P_{oi}(\zeta) \left(\frac{I}{\xi}\right)^{m_o-i} (1+o(1)) = o(1); \quad |\zeta| = r\varepsilon\bar{E}'. \tag{15.9}$$

In both cases for all i $-k_i = \deg P_{ni} - \deg P_{no} < 0$ and

$-1_i = \deg P_{oi} - \deg P_{oo} < 0.$ Besides,

$$\frac{P_{ni}(\zeta)}{P_{no}(\zeta)} = (1 + o(1))A_i \zeta^{-k_i}, \quad i = 1,2\ldots, m_n, \tag{15.9'}$$

$$\frac{P_{oi}(\zeta)}{P_{oo}(\zeta)} = (1 + o(1)) B_i\zeta^{-1_i}, \quad i = 1,2\ldots,m_o, \tag{15.9''}$$

with $A_i,B_i \neq 0$, for all possible i.

We deal below with (15.8) only ((15.9) can be dealt with similarly).  In the neighborhood of $\zeta = \infty$ all the solutions of (15.8) are bounded because of (15.9').  From (15.8) we, consequently, conclude:

$$\left(\frac{I}{|\zeta|}\right)^{m_n} \leq \sum_{i=0}^{n} (1+o(1)) \, |A_i| \, (\zeta)^{-k_i} \left(\frac{I}{|\xi|}\right)^{m_n - i} <$$

$$\leq (1+o(1)) \max_{|\leq i \leq m_n} |A_i| \, |\zeta|^{-\sigma} \sum_{i=1}^{m_n} \left(\frac{I}{|\xi|}\right)^{m_n - i}$$

where $\sigma = \min (k_1, k_2, \ldots, k_n)$.  But $|\xi| = r^{\gamma}$ so that

$$\left(\frac{I}{r^{\gamma}}\right)^{m_n} \leq C' \, \frac{I}{|\zeta|^{\sigma}} \sum_{i=1}^{m_n} (I)^{m_n - i} \leq m_n C; \, I^{m_n - 1}$$

and

$$I < C \frac{r^{\gamma}}{|\zeta|^{\sigma}} \, , \; r\epsilon \overline{\overline{E}} \; . \tag{15.10}$$

We have $|a(r)| = |a_j| = R_j; \; R_j \leq r \leq R_j^{1+\beta}$.  Further $|\zeta| = |a(r) + \xi|$;
$|\xi| = \tau^{\gamma} \leq R^{\gamma(1+\beta)}$.  But $\gamma$ and $\beta$ can be chosen so small that $\gamma(1+\beta) < \min(1, \sigma)$ and then

$$|\zeta| = |a(r) + \xi| = |a(r)| \left(1 + \frac{\xi}{a(r)}\right) = R_j \, (1+o(1)),$$

since $\left|\frac{\xi}{a(r)}\right| \leq \frac{R_j^{\gamma(1+\beta)}}{R_j} \xrightarrow[j \to \infty]{} 0$.  Now from (15.10)

$$I \leq C \frac{R_j^{\gamma(1+\beta)}}{R_j^{\sigma}} = \frac{C}{R_j^{\sigma - \gamma(1+\beta)}} \; 0 \tag{15.11}$$

because $\gamma$ and $\beta$ were picked out satisfying $\sigma - \gamma(1+\beta) < 0$.

But on $\overline{\overline{E}}$ since $r^{\gamma} > 1$ (see 15.5)) the central index $\nu_{a_{j_k}}(r)$ is a non-

decreasing function and

$$I_o(r) = \nu_{a_{j_k}}(R_{j_k} | 1) \leq \nu_{a_{j_k}}(r | t^{\tau}) = I(r | t^{\gamma}, a) \equiv I(r), \; R_{j_k} \leq r \leq R_{j_k}^{1+\beta} \, ,$$

so that

$$\lim_{r \to \infty} I_o(r) = 0$$

which contradicts our assumption (15.5).  Thus $I_o(r) < \infty$ and the theorem is proven.

15.  Proving Theorem 4 in order to verify (4.17) we required the segments $[R_j', R_j'']$ of E to be sufficiently large.  This requirement can be weakened unless we want (15.7) to hold.

Lemma 6.  Let $f(z)$ be a transcendental entire function.  Suppose that $I(r|p,a) \equiv I(r)$ is a g.i on a basic set $E = \bigcup_{j=1}^{\infty} \{[R_j, (1+h)R_j]\}$ with an arbitrary but fixed constant $h > 0$ and tha path.

$$a(r) = a_j; \quad R_j \le r \le (1+h)R_j, \quad j = 1,2,3... \quad .$$

Under these conditions for a given arbitrarily small d:  $0 < d < h$ there is a subset of E

$$E^* = \bigcup_{k=1}^{\infty} \{[(1+d)R_{j_k}, (1+q)R_{j_k}]\}$$

with positive constants d and q:$d < q < h$ such, that either

i)  $\ln M_{a(r)}(p(r)) \ge (1+o(1)) \dfrac{I(r)}{\ln^2 I(r)}, \quad r \varepsilon E^{*\prime}.$       (16.1)

or

ii)  $\ln M_{a(r)}(p(r)) \le -(1+o(1)) \dfrac{I(r)}{\ln^2 I(r)}; \quad r \varepsilon E^{*\prime}.$       (16.2)

Proof.  The proof is essentially the same as of Theorem 4.  Let.

$$R_j^o = \mathrm{Inf}\{r: R_j \le r \le R_j(1+h), M_{a_j}(p(r)) \ge 1\}.$$

a)  Suppose there is a sequence $\{j_k\}$ such that $R_{j_k}^o = (1+\sigma_k)R_{j_k}; \sigma_k \underset{k+\infty}{\to} 0.$  Put

$$\bar{\bar{E}} = \bigcup_{k=1}^{\infty} \{[(1+\sigma_k)R_{j_k}, (1+h)R_{j_k}]\}$$

on $\bar{\bar{E}}$ obviously $\ln M_{a(r)}(p(r)) \ge 0.$

So we come to (4.14) on $\bar{\bar{E}}^*$ in the same way as we have it done in Section 13, that is we get (16.1).

b)   Suppose now that $\displaystyle\lim_{j\to\infty} \frac{R_j^o}{R_j} = 4q'$, $q' > 0$.

Put

$$\bar{\bar{E}} = \bigcup_{j=1}^{\infty} \{[R_j, (1+q)R_j]\} \quad ; \quad 0 < q < q'.$$

Evidently $\ln M_{a(r)}(p(r)) < 0$ on $\bar{\bar{E}}$ and as in Section 13 we obtain (16.2).

17.   Consider now again equation (15.1) with (15.2).  Theorem 5 can be modified in the following manner.

Theorem 6.   Suppose that in (15.1) $\deg P_{oo} = \deg P_{no} \geq \deg P_{ij}$ for all possible $i,j$.  Then each entire solution of (15.1) is of bounded index.

Proof.   Suppose the theorem to be wrong.  Then there is an entire transcendental solution $w(z)$ of (15.1) for which the index $I_o(r)$ is unbounded. Let $\{a_j\}$ with $|a_j| = R_j \uparrow \infty$ be a sequence of points such that $I_o(R_j) = \nu_{a_j}(R_j|1)$ and

$$\overline{\lim_{r\to\infty}}\ I_o(r) = \lim_{j\to\infty} I(R_j) = \infty. \tag{17.1}$$

We define

$$p(r) = \frac{r}{R_j} \ ; \quad R_j \leq r \leq R_j(1+h),\ j = 1,2,\ldots \tag{17.2}$$

where $h>0$ is an arbitrary but fixed number.  Obviously $\omega(r) = \dfrac{rp'(r)}{p(r)} = 1 > 0$.

Now the sequence $\{R_j\}$ can be chosen so rare that

i.   $R_{j+1} > R_j(1+h)$ and

ii.   $2\nu_{a_j}((1+h)R_j|P) < \nu_{a_{j+1}}(r_{j+1}|p)$ . \hfill (17.3)

The function

$$I(r|p,a) = \nu_{a_j}(r|p); \quad R_j \leq r \leq R_j(1+h), \quad j=1,2\ldots$$

is a g.i on the basic set

$$E = \bigcup_{j=1}^{\infty} \{[R_j, (1+h)R_j]\}$$

with the path $a(r) = R_j$ ; $R_j \leq r \leq (1+h)R_j$ ; $j=1,2,3,\ldots$  .

The same considerations as the ones proving Theorem 5 bring us to equation (15.7).  In the present case the lower bound of the variable $|\zeta| = p(r)$ on the segment $[R_j, (1+h)R_j]$ equals 1 the upper $-1+h$.  Under the conditions of the theorem the ratios $\dfrac{P_{k1}(\zeta)}{P_{no}(\zeta)}$ and $\dfrac{P_{k1}(\zeta)}{P_{oo}(\zeta)}$ have finite upper bounds independent of $k,l$.  Now in view of Lemma 6 and the boundedness of the ratios above we get from (15.7) in case $w(\zeta) \underset{r\to\infty}{\to} \infty$ on $\bar{E}'$  Equation (15.8); in case of $w(\zeta) \underset{r\to\infty}{\to} 0$ on $\bar{E}'$ - Equation (15.9).  Suppose $w(\zeta) \to \infty$ on $\bar{E}'$.  Then we obtain from (15.8)

$$\left(\frac{I}{|\xi|}\right)^{\frac{m_n}{n}} \leqslant C' \sum_{i=0}^{m_n} \left(\frac{I}{|\xi|}\right)^{\frac{m_n-i}{n}} < C'' \left(\frac{I}{|\xi|}\right)^{\frac{m_n-1}{n}} ==> I < C < \infty ,$$

since $1 < |\xi| < 1+h$, in contrary with our assumption $I(r) \to \infty$, $r \varepsilon E$.  To the same conclusion and in the same way we also come if $w(\zeta) \underset{|\zeta|\to\infty}{\to} 0, |\xi| \varepsilon \bar{E}$.

But $I_o(r) < I(r)$ on $\bar{E}'$.  Thus $I_o(r) < \infty$.

Hence Theorem 6 is correct.

18.   Consider now the idfferential equation

$$\sum_{j=0}^{n-1} Q_j(z,w')w^{n-j} = g(z), \qquad (18.1)$$

where $g(z)$ is an entire transcendental function of bounded index on the complex plane.  (Note that in the previous section $g(z)$ was considered a polynomial).  We deal below with a more general equation

$$\sum_{j=0}^{n-1} P_j(z,\frac{w'}{w})w^{m-j} = g(z) \qquad (18.2)$$

(which, obviously includes also Equation (18.1) where

$$P_j(z,\eta) = \sum_{i=0}^{m_j} P_{ji}(z)\eta^{m_j-i} \qquad (18.3)$$

with polynomials $P_{ji}(z)$ and an entire transcendental function $g(z)$ of bounded index: $I_o(r,g) < \infty$.  The following theorem holds.

Theorem 7.  Suppose that in (18.1) - (18.2) either $\qquad\qquad$ (18.4)

a)  $\deg P_{oo} > \deg P_{ok}$, $k=1,2,\ldots m_o$; $\deg P_{n-1,0} > \deg P_{n-1,k}$, $k-1,2.,m_{n-1}$

or

b) $\deg P_{oo} = \deg P_{n-1} \geq \deg P_{ij}$ for all possible i,j.                    (18.5)

Then every entire solution of (18.1) is of bounded index.

Proof.  We will prove here only case b).  Case a) can be dealt with analogously by a suitable choice of the function p(r) as in Section 16.  Suppose that our theorem is wrong so that there is a transcendental entire solution w(z) of

(18.2) with $\overline{\lim_{r \to \infty}} \ I_o(r) = \infty$.  As in Section 17 proving Theorem 6, let $R_j \uparrow \infty$

be a sequence such that

$$\lim_{r \to \infty} I_o(r) = \lim I_o(R_j) = \infty \tag{18.6}$$

and $a_j : \ |a_j| = R_j$ the sequence of points on which

$$I_o(R_j) = \nu_{a_j}(R_j|1) = \max_{|a|=R_j} \nu_a(R_j|1). \tag{18.7}$$

Take

$$p(r) = \frac{r}{R_j}, \ R_j \leq r \leq (1+h)R_j; \ j = 1,2,3\ldots,\omega(r)\equiv1 \tag{18.8}$$

and the sequence $\{R_j\}$ so rare that

$$2\nu_{a_j}((1+h)R_j|1) < \nu_{a_{j+1}}(R_{j+1}|1).$$

Now on the basic set

$$E = \bigcup_{j=1}^{\infty} \ \{[R_j, \ (1+h)R_j]\} \quad \text{with the path}$$

$$a(r) = a_j \ ; \ R_j \leq r \leq (1+h)R_j, \ j=1,2,3,.. \tag{18.9}$$

the function

$$I(r) \equiv I(r|p,a) = \nu_{a_j}(r|p), \ R_j \leq \tau \leq (1+h)R_j, \ j=1,2,3\ldots \tag{18.10}$$

is a g·i on E. (4.8) shows that at each point $\zeta$ on the circle,

$$|\xi| = |\zeta - a(r)| = p(r) \ (\xi = \zeta - a(r)) \text{ where } w(\zeta) = M_{a(r)}(p(r)), \ r\varepsilon E'$$

$$\left| (\xi\frac{d}{d})^n \ln(a(r) + \xi)\right| < n![I(r)\ln^{3+\alpha}I(r)]^{\frac{n}{2}} \ln I(r), \ n=2,3,4.. \ . \tag{18.11}$$

Besides, according to (4.12) on E'

$$w^{(k)}(\zeta) = (1+o(1)) \ (\frac{I}{\xi})^k w(\zeta), \ k=1,2,3\ldots \ . \tag{18.12}$$

SH. STRELITZ

19.  We go on with the proof of Theorem 7.  The sum in (18.2) is a sum of terms of the form $Q = P(z) \left(\frac{w'}{w}\right)^P w^q$.

It can be easily proven by induction that

$$\left(\frac{w'}{w}\right)^{(k)} = \frac{1}{z^k} \sum_{j=1}^{k} A_j^{(k)} \left(z\frac{d}{dz}\right)^j \ln w \tag{19.1}$$

with appropriate constants $A_j^{(k)}$.

Now in view of (18.11) and (18.12) for $z = \zeta = \xi + a(r)$

$$\left|\left(\frac{w'}{w}\right)^{(k)}\right| \leq \frac{1}{|\xi|^k} \sum_{j=1}^{k} \left|A_j^{(k)}\right| \; j! \; [I(r)\ln^{3+\alpha} I(r)]^{\frac{j}{2}} \ln I(r) \quad <$$

$$< \frac{C_k}{|\xi|^k} [I(r)\ln^{3+\alpha} I(r)]^{\frac{k}{2}} \ln I(r) \leq C_k [I(r)\ln^{3+\alpha} I(r)]^{\frac{k}{2}} \ln I(r) \tag{19.2}$$

because in our case $1 \leq |\xi| \leq 1+h$ and $I(r) \to \infty$.

We have

$$Q' = P'\left(\frac{w'}{w}\right)^P w^q = P'\left(\frac{w'}{w}\right)^P w^q + p\, P\left(\frac{w'}{w}\right)' w^q + q\, P\left(\frac{w'}{w}\right)^{p+1} w^q. \tag{19.3}$$

At the points $\zeta$ with $r\varepsilon E'$ according to (18.12) we have:

$$Q'(\zeta) = (1+o(1))P'\left(\frac{I}{\xi}\right)^P w^q + (1+o(1))pP\left(\frac{I}{\xi}\right)^{p-1}\left(\frac{w'}{w}\right)' w^q + (1+o(1))qP\left(\frac{I}{\xi}\right)^{p+1}w^q,$$

$$I \equiv I(r). \tag{19.4}$$

By (19.2)

$$\left|\frac{I}{\xi}\right|^{p-1} \left|\left(\frac{w'}{w}\right)'\right| < C_1 \frac{I^{p-1}}{|\xi|} [I\ln^{3+\alpha} I]^{\frac{1}{2}} \ln I < C_1 I^{p-\frac{1}{2}} \ln^{\frac{5+\alpha}{2}} I$$

(since $|\xi| \geq 1$).  With respect to this inequality we get from (19.4):

$$Q'(\zeta) = (1+o(1))qP(\zeta) \left(\frac{I}{\xi}\right)^{p-1} w^q. \tag{19.5}$$

It is not difficult to show by induction that

$$Q^{(m)} = \left\{\Sigma A_{i_o i_j \ldots i_m} \left(\frac{w'}{w}\right)^{i_1} \left[\left(\frac{w'}{w}\right)'\right]^{i_2} \ldots \left[\left(\frac{w'}{w}\right)^{(m)}\right]^{i_m} p^{(i_o)} + q^m \left(\frac{w'}{w}\right)^{p+m} p\right\} w^q$$

where $i_o + i_1 + 2i_2 + \ldots + mi_m = p+m$; $\; i_1 \neq p+m$ $\tag{19.6}$

with certain constants $A_{i_o i_1 \ldots i_m}$.

At $\zeta$ with regard to 919.2) and (18.12)

$$T_{i_1 \ldots i_m} = \left| \left(\frac{w'}{w}\right)^{i_1} \left[\left(\frac{w'}{w}\right)'\right]^{i_2} \ldots \left[\left(\frac{w'}{w}\right)^{(m)}\right]^{i_m} \right| <$$

$$< B_{i_1 i_2 \ldots i_m} I^{i_1 + 2/2 i_2 + 3/2 i_3 + \ldots + \frac{m}{2} i_m} \ln^{(1+\alpha) \sum\limits_{j=2}^{m} \frac{j}{2} + 1} I$$

with $B_{i_1 i_2 i_m} = $ Cons. But $i_1 \le p+m-1$ because the term with $i_1 = p+m$ is detached in (19.6).

Therefore

$$i_1 + \tfrac{1}{2} \sum_{j=2}^{m} j i_j \le \frac{p+m+i_1}{2} \le p+m-\tfrac{1}{2},$$

so that

$$T_{i_1 \ldots i_m} < B_m I^{p+m-\frac{1}{2}} \ln^{(3+\alpha)(p+m+\frac{1}{2})} 1.$$

Now (19.6) gives us

$$\left| \Sigma A_{i_0 i_1 \ldots i_m} P^{(i_0)} \prod_{k=1}^{m} \left[\left(\frac{w'}{w}\right)^{(k)}\right]^{i_k} \right| <$$

$$< \left( \Sigma \left| A_{i_0 i_1 \ldots i_m} \right| B_{i_1 i_2 \ldots i_m} |P| \right) I^{p+m-\frac{1}{2}} \ln^{(3+\alpha)(p+m+\frac{1}{2})}$$

where $i_0 + \sum\limits_{k=1}^{m} k i_k = p+m$; $i_1 \ne p+m$. Applying this inequality to (19.6) we obtain:

$$Q^{(m)}(\zeta) = (1+o(1)) \, a^m \left(\frac{I}{\xi}\right)^{p+m} p(\zeta) \, w^q(\zeta). \tag{19.7}$$

20. Each polynomial $P_j(z, \frac{w'}{w}) w^{n-m_j}$ is a finite sum of terms of the form $Q$ considered in the previous section. According to (18.3) and (19.7)

$$\frac{d^m}{dz^m} \left[ P_j(z, \frac{w'}{w}) \, w^{n-j} \right]_{z=\zeta} =$$

$$= \sum_{i=0}^{m} (1+o(1)) (n-j)^m P_{ji}(\zeta) \left(\frac{I}{\xi}\right)^{mj-i} w^{n-j}(\zeta). \tag{20.1}$$

By differentiation of Equation (18.2) successively N times at the points $\zeta: |\zeta| \in E'$ according to (20.1) we will have:

$$\sum_{j=0}^{n-1} \sum_{i=0}^{m_j} (1+o(1)) P_{ji}(\zeta) \left(\frac{I}{\xi}\right)^{m_j-i} (n-j)^k w^{n-j}(\zeta) = g^{(k)}(\zeta), \; k=0,1,2\ldots,N. \tag{20.2}$$

200                                 SH. STRELITZ

Lemma 6 shows that on a certain subset

$$\bar{E} = \bigcup_{k=1}^{\infty} \left\{ \left[ (1+d)R_{j_k} , (1+q)R_{j_k} \right] \right\} ; \qquad 0 < d < q < h$$

either

$$\frac{I^S(r)}{M_{a(r)}(p(r))} \quad \underset{\substack{r \to \infty \\ r\varepsilon\bar{E}'}}{\longrightarrow} \quad 0 \tag{20.3}$$

and $|w(\zeta)| \to \infty$

or

$$I^S(r)\, M_{a(r)}(p(r)) \quad \underset{\substack{r \to \infty \\ r\varepsilon\bar{E}'}}{\longrightarrow} \quad 0 \tag{20.4}$$

and $|w(\zeta)| \to 0.$

Assume first the case (20.3) to take place. Then from (20.2) it follows.

$$\sum_{i=0}^{m_o} (1+o(1)) \frac{P_{oi}(\zeta)}{P_{oo}(\zeta)} \left(\frac{I}{\xi}\right)^{m_o-i+k} n^k w^n(\zeta) +$$

$$+ w^q(\zeta) \sum_{i=0}^{n=1} \sum_{i=0}^{m_j} (1+o(1)) \frac{P_{ij}(\zeta)}{P_{oo}(\zeta)} \left(\frac{I}{\xi}\right)^{m_j-i+k} (n-j)^k w^{-j}(\zeta) = \frac{g^{(k)}(\zeta)}{P_{oo}(\zeta)}. \tag{20.5}$$

But $\deg \dfrac{P_{oi}}{P_{oo}} = \deg P_{oi} = \deg P_{oo} \leq 0$, so that for a certain sequence

$\zeta = \zeta_t$; $t = 1,2,3,\ldots$ since $I \to \infty$ .

$$\sum_{i=0}^{m_o} (1+o(1)) \frac{P_{oi}(\zeta)}{P_{oo}(\zeta)} \left(\frac{I}{\xi}\right)^{m_o-i+k} = (1+o(1)) \left(\frac{I}{\xi}\right)^{m_o+k} . \tag{20.6}$$

On the other hand in view of (20.3) the double sum in (20.5) tends to zero as $\zeta_i$ tending to $\infty$. Thus with respect to (20.5) we obtain from (20.6)

$$(1+o(1)) \left(\frac{I}{\xi}\right)^{m_o+k} n^k w^n(\zeta) = \frac{g^{(k)}(\zeta)}{P_{oo}(\zeta)} ; \quad k=0,1,2,\ldots,N+1 .$$

Now (14.2) (substituting there g(z) for f(z) gives us:

$$\sum_{i=1}^{N} (1+o(1)) \left|\frac{I}{\xi}\right|^{m_o+k} \frac{n^k}{k!} \left|w^n(\zeta)\right| \geq$$

$$\geq \frac{(1+o(1)) \dfrac{I}{\xi}^{m_o+n+1}}{(N+1)!} w^n(\zeta)$$

so that

$$\sum_{k=1}^{N} \left(\frac{n^k}{k!}\right) \geq \frac{(1+o(1))}{(N+1)!} \left|\frac{I}{\xi}\right|.$$

But this inequality is impossible because of the assumption $I(r) \xrightarrow[i\to\infty]{} \infty$;

$$r_i = |\zeta_i| .$$

Hence (20.3) is wrong on $\bar{E}'$.

Similar considerations, which we omit here and leave the detailed computation for the reader, show that (20.4) is impossible too. Thus $I(r)$ is bounded and so is $I_o(r)$.

The theorem is proven.

Remark. The theorem is evidently correct if all the $P_{ij}(z)$ are constants.

§6.  On the Growth of the Index $I(r)$ of Entire Transcendental Solutions of

First Order Algebraic Differential Equations.

21.  Consider the differential equation

$$F_o(z,w,w') \equiv \sum_{i+j=0}^{n} Q_{ij}(z)\, w^i w'^j = 0$$

where all the $Q_{ij}(z)$ are polynomials.  For conveniency we rewrite the last equation as in the previous paragraph in the form

$$\sum_{j=0}^{n} P_j\left(z,\frac{w'}{w}\right) w^{n-j} = 0 \tag{21.1}$$

with

$$P_j(z,\eta) = \sum_{i=0}^{m_j} P_{ji}(z)\eta^{m_j-i}$$

and

$$P_j(z) = A_{ji}(1+o(1))z^{\alpha_{ji}} .$$

Let w(z) be a transcendental entire solution of (21.1).  Our purpose it o
estimate the growth of the index

$$I_\alpha(r) = \max_{|a|=r} \nu_a(r|t^\alpha)$$

with $0 < \alpha < 1$ of the function $w(z)$.

Assume

$$\varlimsup_{r\to\infty} I_\alpha(r) = \infty .$$

Under this assumption there is a sequence $R_j \uparrow \infty$ such that

$$\varlimsup_{r\to\infty} \frac{\ln I_\alpha(r)}{\ln r} = \lim_{j\to\infty} \frac{\ln I_\alpha(R_j)}{\ln R_j} \tag{21.2}$$

and a corresponding sequence of points $a_j$ on the complex plane such that

$$I_\alpha(R_j) = \nu_{a_j}(R_j|t^\alpha); \quad |a_j| = R_j \uparrow \infty. \tag{21.3}$$

Our next step is to construct a g.i. $I(r|t^\alpha,a)$ on a certain basic set

$$E = \bigcup_{j=1}^{\infty} \{[R_j, R_j^{1+\beta}]\},$$

with an arbitrary but fixed $\beta > 0$ and a path

$$a(r)=a_j; \quad r\epsilon[R_j, R_j^{1+\beta}], \quad j=1,2,3\dots . \tag{21.4}$$

Besides we assume

$$\{[R_j, R_j^{1+\beta}]\} \cap \{[R_k, R_k^{1+\beta}]\} = \emptyset, \quad j\neq k .$$

In our case we choose $p(t) = t^\alpha$ , so that $\omega(t) \equiv \alpha > 0$.

We define

$$I(r) \equiv I(r|t^\alpha,a) = \nu_{a_j}(r|t^\alpha); \quad r\epsilon [R_j, R_j^{1+\beta}], \quad j=1,2,3\dots$$

and choose the sequence $\{R_j\}$ so rare that

$$2\nu_{a_j}(R_j^{1+\beta}|t^\alpha) < \nu_{a_{j+1}}(R_{j+1}|t^\alpha), \quad j = 1,2,3\dots .$$

Evidently

$$I_\alpha(r) \leq I(r); \quad r\epsilon E \tag{21.5}.$$

22.  By (4.13)

$$w'(\zeta) = (1+o(1)) \frac{I}{\xi} w(\zeta); \quad I \equiv I(r), \quad \xi = \zeta - a(r), \quad r\varepsilon E \tag{22.1}$$

where $\zeta$ is a point on the circle $|\xi| = |\zeta - a(r)| = r^{\alpha}$ with $|w(\zeta)| = M_{a_j}(r^{\alpha})$.

Insert now in (21.1) $z = \zeta$. Then according to (22.1)

$$\sum_{j=0}^{n} \sum_{i=0}^{m_j} (1+o(1)) A_{ji} \zeta^{\alpha ji} \left(\frac{I}{\xi}\right)^{m_j-i} w^{n-j}(\zeta) = 0. \tag{22.2}$$

We now come back to Equation (21.1) and its entire solution $w(z)$. Let $E^*$ be one of the subsets $\bar{E}$ or $\bar{\bar{E}}$ considered in Theorem 5.

1. Suppose first $E^* = \bar{E}$, so that

$$\ln|w(\zeta)| = \ln M_{a(r)}(r^{\alpha}) > (1+(o)) \frac{I(r)}{\ln^2 I(r)}.$$

Now we rewrite (22.2) as follows

$$\sum_{i=0}^{m_o} (1+o(1)) A_{oi} \zeta^{\alpha oi} \left(\frac{I}{\xi}\right)^{m_o-i} =$$

$$= -\sum_{n=1}^{n} \sum_{i=0}^{m_j} (1+o(1)) A_{ji} \zeta^{\alpha ji} \left(\frac{I}{\xi}\right)^{m_j-i} w^{-j}(\zeta). \tag{22.3}$$

By Theorem 4 for every pair constants, $s \geq 0$ and $t \geq 0$ on $r\varepsilon \bar{E}'$

$$\frac{|\zeta| \tau |\frac{I}{\xi}|^s}{M_{a(r)}(r^{\alpha})} < \frac{(r+r^{\alpha})^t r^{\alpha s} I^s}{M_{a(r)}^{\frac{1}{2}}(r^{\alpha}) \exp\{(1+o(1))\frac{\sqrt{I}}{\ln I}\}} \xrightarrow[r\varepsilon \bar{E}']{r\to\infty} 0.^0. \tag{22.4}$$

Hence (22.2) is an algebraic equation

$$\sum_{i=0}^{m_o} (1+o(1)) A_{oi} \zeta^{\alpha oi} \left(\frac{I}{\xi}\right)^{m_o-i} = o(1).$$

For conveniency we put $A_{oi} = A_i$ and obtain

$$\sum_{i=0}^{m_o} (1+o(1)) A_i \zeta^{a_i} \left(\frac{I}{\xi}\right)^{m_o-i} = o(1), \quad r\varepsilon \bar{E}'.. \tag{22.5}$$

As it is well known (see for example 12 ) Equation (22.5) has $m_o$ solutions of the form

$$I = (1+o(1)) B_j \zeta^{\alpha_j} \xi, \quad j=1,2,\ldots,m_o \;. \tag{22.6}$$

Suppose

204                         SH. STRELITZ

$$\bar{E} = \bigcup_{j=1}^{\infty} \{[\bar{R}_k, \bar{R}_k^{\,1+\frac{\beta}{2}}]\}, \quad R_k = R_{j_k}^{\,1+\frac{\beta}{4}}, \quad k = 1,2,3,\ldots .$$

We restrict ourselves now on the subset $\bar{E}* = \bigcup^{\infty} \{[\bar{R}_k, (1+h)\bar{R}_k]\}$ with a fixed

$h > 0$. We have $|\zeta| = r^{\alpha}$ and $a(r) = R_{j_k}$; $\bar{R}_k \leq r \leq (1+h)\bar{R}_k$. Then

$|\zeta| = |a(r) + \xi| = (1+o(1))R_{j_k}$ and $\ln r = (1+o(1))(1+\frac{\beta}{4})\ln R_{j_k}$ if $\beta > 0$ is

chosen so small that $\alpha(1+\frac{\beta}{4}) < 1$ and $|\xi| = r^{\alpha} < (1+h)r_{j_k}^{\,\alpha(1+\frac{\beta}{4})}$. Consequently

$$\overline{\lim_{\substack{r\to\infty \\ r\in\bar{E}*'}}} \frac{\ln I(r)}{\ln r} = \frac{\sigma + \alpha(1+\frac{\beta}{4})}{1 + \frac{\beta}{4}} \tag{22.7}$$

where $\sigma > 0$ is one of the exponents in (22.6). But

$$I_{\alpha}(R_{j_k}) \leq I((1+h)\bar{R}_k) = I((1+h)R_{j_k}^{\,1+\frac{\beta}{4}}).$$

Therefore

$$\overline{\lim_{r\to\infty}} \frac{\ln I_{\alpha}(r)}{\ln r} = \lim_{k\to\infty} \frac{\ln I_{\alpha}(R_{j_k})}{\ln R_{j_k}} \leq$$

$$\leq \overline{\lim_{\substack{r\to\infty \\ r\in\bar{E}*'}}} \frac{\ln I(r)}{\ln r} = \frac{\sigma + \alpha(1+\frac{\beta}{4})}{1+\frac{\beta}{4}} .$$

Since $\beta < 0$ is an arbitrary number then

$$\overline{\lim_{r\to\infty}} \frac{\ln I_{\alpha}(r)}{\ln r} \leq \sigma + \alpha . \tag{22.7}$$

It is known (see [2]) that if $w(z)$ is an entire function of order $\rho$ then

$$\rho + \alpha - 1 \leq \overline{\lim_{r\to\infty}} \frac{\ln I_{\alpha}'(r)}{\ln r} < \rho \tag{22.8}$$

Theorem 4 shows that $\sigma + \alpha \leq \rho$ (there is in our case

$$\frac{I(r)}{\ln^2 I(r)} \leq \ln_{a(r)} M(r^{\alpha}) \leq \ln M(R_{j_k} + R_{j_k}^{\,\alpha(1+\frac{\beta}{4})}(1+h)^{\alpha}), \quad R_{j_k}^{\,1+\frac{\beta}{4}} \leq r \leq R_{j_k}^{\,1+\frac{\beta}{4}}(1+h),$$

where $M(r) = \max\limits_{|z|=r} |f(z)|)$. Thus $\rho + \alpha - 1 \leq \sigma + \alpha \leq \rho$.

If there is no $\sigma_j$ in (22.6) such that $\sigma + \alpha = 1 < \sigma_j + \alpha$, then

$$\overline{\lim_{r \to \infty}} \frac{\ln I_\alpha(r)}{\ln r} = \rho + \alpha - 1.$$

2. There may take place the case $E^* = \overline{\overline{E}}$ of Theorem 4, so that $\ln M_{a(r)}(r^\alpha) < 0$
on $\overline{\overline{E}}$. In this case instead of (22.4) we will have on $\overline{\overline{E}}'$

$$|\varsigma|^t \left(\frac{I}{|\xi|}\right)^s |w(\zeta)| \xrightarrow[\substack{r \to \infty \\ r \varepsilon \overline{\overline{E}}'}]{} 0. \tag{22.9}$$

We infere from (22.2) that

$$\sum_{i=0}^{m_n} (1+o(1))\, A^*_i \varsigma^{\alpha^*_i} \left(\frac{I}{\xi}\right)^{m_n - i} = o(1), \tag{22.10}$$

where $A^*_i = A_{ni}$ and $\alpha^*_i = \alpha_{ni}$. From (22.10) $m_n$ solutions

$$I = (1+o(1))\, B_j^* \sigma^{\sigma_j} \xi, \quad j = 1, 2, \ldots, m_n,$$

follow. Hence in this case we obtain, too, that

$$\overline{\lim_{r \to \infty}} \frac{\ln I_\alpha(r)}{\ln r} \leq \sigma + \alpha,$$

where this time $\sigma$ coincides with one of the positive exponents $\sigma^*_j$. Further
$\sigma + \alpha < \rho$, because according to Theorem 4

$$\frac{(1+o(1))I(r)}{\ln^2 I(r)} < -\ln M_{a(r)}(r^\alpha). \tag{22.11}$$

It is well known that in case of an entire function of order $\rho$ for every
z on the complex plane outside maybe a set of discs with a finite sum of radii
$|f(z)| > A\exp(-r^{\rho + \varepsilon})$ where $\varepsilon > 0$ is an arbitrary fixed number and $A = A(\varepsilon) = Const$.

Therefore in view of (22.11)

$$\frac{(1+o(1))I(r)}{\ln^2 I(r)} < (R_{j_k} + R_{j_k}^{\alpha(1+\frac{\beta}{8})} (1+h)^\alpha)^{\rho + \varepsilon}$$

for $R_{j_k}^{1+\frac{\beta}{8}} \leq r \leq (1+h) R_{j_k}^{1+\frac{\beta}{8}}$ and henceforth $\sigma + \alpha \leq \rho$. Consequently if there is

no $\sigma^*_j$ such that $\sigma^*_j + \alpha > \rho + \alpha - 1$ then

$$\overline{\lim_{r \to \infty}} \; \frac{\ln I_\alpha(r)}{\ln r} = \rho + \alpha - 1. \tag{22.12}$$

Note that if $\alpha$ is near enough to 1: $1 > \alpha \gtrsim \alpha_o$ with an $\alpha_o$ sufficiently large then (22.12) holds always. Indeed if $\rho + \alpha - 1 < \sigma_j + \alpha$ then $\sigma_j > \rho - 1$ and for $\alpha$ near to $1, \sigma_j + \alpha > \rho$ (the same is evidently true also for $\sigma_j^*$).

§7.   Some other expamples.

23.   In this paragraph we have in mind to show how the previously built Wiman-Valiron theory can be applied to higher order differential equations. In this section we use our theory to estimate the growth of the index $I_o(r)$ of holomorphic solutions, in an angle, of a linear differential equation.

Consider a linear differential equation

$$y^{(n)} + a_1(z) y^{(n-1)} + \ldots + a_n(z) y = 0 \tag{23.1}$$

with analytic coefficients $a_j(z)$, $j = 1, 2 \ldots, n$ in an angle $A' : |\text{ang } z| < \theta'$. We assume the following asymptotic expression

$$a_j(z) = (1 + o(1)) A_j \, z^{\alpha_j}, \quad j = 1, 2, \ldots n, \quad o(1) \xrightarrow[z \to \infty]{} 0 \tag{23.2}$$

with complex fixed $\alpha_j$ and $o(1) \to 0$ uniformly in each closed angle

$A: |\text{ang } z| < \theta_o < \theta'$ belonging to $A'$.

Let $y(z)$ be a solution of Equation (23.2) in $A'$. We are going to evaluate the index

$$I_o(r) = \max_{\substack{|a|=r \\ a \varepsilon A}} \nu_a(r|1)$$

where $A = \{z : |\text{ang } z| \leq \theta_o < \theta'\}$.

Suppose $I_o(r)$ to be unbounded in $A$. We use below the g.i $I(r|p,a)$ constructed in section 18 (see (18.7)) and further up to (18.10). Then

$$\overline{\lim_{r \to \infty}} \; \frac{\ln I_o(r)}{\ln r} < \overline{\lim_{\substack{r \to \infty \\ r \varepsilon E}}} \; \frac{\ln I(r|p,a)}{\ln r} \; . \tag{23.3}$$

We apply to this g.i our Theorem 3. At the points $\zeta$ on the circles $|z - a(r)| = p(r)$ where

$$\left| y(\zeta) \right| = \left| y(\xi + a(r)) \right| = M_{a(r)}(p(r)); \quad \zeta = a(r) + \xi$$

we, according to Theorem 3, obtain

$$y^{(n)}(\zeta) = (1+o(1)) \frac{I^n(r)}{\xi^n} y(\zeta), \quad n=1,2,3,\ldots, \ r\epsilon E' \ . \tag{23.4}$$

Now from (23.1) it follows

$$(\frac{I}{\xi})^n + (1 + o(1))\, a_n(\zeta)\, (\frac{I}{\xi})^{n-i} +\ldots+ (1 + o(1))\, a_n(z) = 0$$

with $\left|\xi\right| = p(r)$, $1 < \left|\xi\right| < 1+h$ and $I \equiv I(r) \equiv I(r|p,a)$.

But $a_j(z) = (1+o(1))\, A_j\, z^{\alpha_j}$. Consequently

$$(\frac{I}{\xi})^n + \sum_{j=1}^{n} + (1+o(1))\, A_j \zeta^{\alpha_j} (\frac{I}{\xi})^{n-j} = 0. \tag{23.5}$$

We have here an algebraic equation for $\frac{I}{\xi}$ with complex exponents

$\alpha_j = \sigma_j + i\, \tau_j$, $j = 1,2,\ldots,n$, where $\sigma_j$ and $\tau_j$ are real numbers. Denote

$A_j = A_j^* e^{i\alpha_j}$ and $\zeta = r.e^{i\phi}$. Thus $\zeta^{\alpha_j} = \zeta^{\sigma_j + i\tau_j} = r^{\sigma_j + i\tau_j}.e^{i\phi(\sigma_j + i\tau_j)} =$

$= r^{\sigma_j} e^{-\phi\tau_j}.e^{i(\phi\sigma_j + \tau_j \ln r)}$. Then (23.5) shows that

$$(\frac{I}{\xi})^n + \sum_{j=1}^{n} (1+o(1)) A_j^* e^{i(\alpha_j + \phi\sigma_j \tau_j \ln r)}\, e^{-\phi\tau_j}\, r^{\sigma_j} (\frac{I}{\xi})^{n-j} = 0. \tag{23.6}$$

Consider now the equation

$$v^n + \sum_{j=1}^{n} r^{\sigma_j} v^{n-j} = 0. \tag{23.7}$$

There exists n solution of (23.7)

$$v = (1+o(1)) B_k r^{\omega_k}, \quad k = 1,2,\ldots,n$$

with real $\omega_k$. If we put in (23.7) $v = u r^{\omega_k}$ we will find

$$u^n + \sum_{j=1}^{n} r^{\sigma_j - \omega_k j} u^{n-j} = 0.$$

Among the numbers $\{\sigma_j - \omega_k j\}_{j=1}^{n}$ with a fixed k there are at least two equal.

Say

$$\sigma_{j_p} - \omega_k j_p = \omega, j_p < j_{p+1}, \ p = 1,2\ldots, \ s_k \ \text{and} \ \sigma_{j_q} - \omega_k j_q < \omega \ \text{if} \ q \neq p.$$

(see for example [11], p. 40). By the transform $\dfrac{1}{\zeta} = ur^{\omega_k}$ we get from (23.6)

$$u^n + \sum_{j=1}^{n} (1+o(1))A_j^* e^{i(\alpha_j + \phi\sigma_j + \tau_j \ln r)} e^{-\phi\tau_j} r^{\sigma_j + \omega_k} u^{n-j} = 0.$$

Divide now the last equality on $r^{\omega}$ with the $\omega$ found above. Then since $\sigma_{j_q} - \omega_k j_q - \omega < 0$, $q \neq p$, we will have

$$\sum_{p=1}^{s} (1+o(1))A_j^* e^{i(\alpha_{j_p} + \phi\sigma_{j_p} + \tau_{j_p} \ln r)} e^{-\phi\tau j_p} u^{j_s - j_p} = o(1). \tag{23.8}$$

But $\rho$ is bounded in A so that (23.8) is an equation with bounded coefficients. As inference we now get that all the solutions of (23.8) are bounded with a lower bound different from zero.

Thus we obtained for this $\omega_k$, $j_{p_{s_k}} - j_{p_1}$ solutions of equation (23.6)

$$\frac{1}{\zeta} = D_m(\zeta) \, r^{\omega_k}, \quad r \varepsilon \overline{E}'$$

with two constants $0 < \beta_1$ and $\beta_2 < \infty$, that $0 \leq \beta_1 < |D_m(\zeta)| \leq \beta_2 < \infty$.

Note that

$$\max_k \omega_k = \max_{1 \leq j \leq n} \frac{\sigma_j}{j} = \rho.$$

Hence

$$I(r) \leq D^* r^{\rho}$$

where $D^*$ is a constant. From (23.3) it follows $I_o(r) \leq D^* r^{\rho}$, that is: $I_o(r)$ is no more then of order $\rho$ and normal type (we define the order as

$$\varlimsup_{r \to \infty} \frac{\ln^+ \ln^+ I_o(r)}{\ln r} = \rho^* \quad \text{and the type as} \quad \varlimsup_{r \to \infty} \frac{\ln^+ I_o(r)}{r^{\rho^*}}.)$$

Obviously the solution must be of bounded index in A too, so that $I_o(r) < \infty$.

It is not difficult to formulate assertions on the boundedness of the index $I_o(r)$ of all soltuions of equation (23.1) in A analogous to those, obtained in [10] and in §5 of the present paper for entire solutions. The same can be said also about solutions of bounded index of the equation.

$$y^n + \sum_{j=1}^{n} a_j(z) y^{(n-j)} = g(z),$$

where $g(z)$ is a function of bounded index in $A$. We will not consider these problems, the solution of which is almost evident.

24. We would like to demonstrate some other possible results using the same technique as above in the previous section.

Consider equation (23.1) once more, where this time all the $a_j(z)$ are rational functions, so that in (23.2) all the $\alpha_j$ are entire numbers. Besides of (23.1) we consider also the characteristic equation

$$u^{(n)} + (1+o(1))A_1 z^{\alpha_1} u^{(n-1)} + \ldots + (1+o(1))A_n z^{\alpha_n} = 0. \tag{24.1}$$

Let

$$u = (1+o(1))B_k z^{\beta_k}, \quad k = 1,2,\ldots n \tag{24.2}$$

be all the solutions of (24.1) in the neighborhood of $z = \infty$.

Proposition 1. Suppose all the $\beta_k$, $k=1,2..,m$ to satisfy the inequalities $\beta_k > 1$, $k=1,2,\ldots,m$ and all the other $\beta_k \leq 0$; $k = m+1, m+z,\ldots,m$. Then all the entire solutions of order $\leq 1$ (if, of course, they exist) are of bounded index: $I_o(r) < \infty$.

Proof. Suppose that $w(z)$ is an entire transcendental solution of (24.1) of order $\rho \leq 1$ and unbounded index $I_o(r)$. As in the previous section we construct a g.i $I(r|p,a)$ on a certain basic set $E = \bigcup_{j=1}^{\infty} [R_j, R_j(1+h)]$ like we have done it in Section 19 satisfying the inequality $\varlimsup_{r \to \infty} I_o(r) \leq \varlimsup_{\substack{r \to \infty \\ r \in E'}} I(r|p,a)$.

Using (23.4) we again come to Equation (23.5), where this time $\alpha_j$ are entire numbers. All the solutions of (23.5) in the neighborhood of $z = \infty$ are given by (24.2), where $u$ has to be replaced by $\frac{I}{\xi}$. But for an entire function the number

$$\varliminf_{r \to \infty} \frac{\ln I(r)}{\ln r} \leq \rho \leq 1 \tag{24.3}$$

(see section 22). In view of the condition of our proposition either $\beta_k > 1$ in (24.2) or $\beta_k \leq 0$, that is it should be $I(r) < C < \infty$ thereby too, $I_o(r) < C < \infty$, which contradicts our assumption that $I_o(r)$ is unbounded.

25.  The assertion of this section will be proven using the technique of the previous section and section 18.  Consider a differential equation

$$a_0(z)y^{(n)} + a_1(z)y^{(n-1)} + \ldots + a_n(z)y = g(z) \tag{25.1}$$

where all the $a_j(z)$ are polynomials and $g(z)$ is an entire transcendental function of order $\rho \leq 1$ and bounded index $I_0(r)$.  Let asymptotically

$$a_k(z) = (1+o(1))\, \overset{\circ}{A_k} z^{\overset{\circ}{\alpha}_k}, \quad k = 0,1,2,\ldots, n \tag{25.2}$$

and

$$\frac{a_k(z)}{a_0(z)} = (1+o(1))\, A_k z^{\alpha_k}, \quad k = 0,1,\ldots, n$$

in the neighborhood of $z = \infty$.  Suppose further that (24.2) gives all the solutions of the characteristic equation (24.1), where $\beta_k > 1$, $k = 1,2,\ldots,m$ and $\beta_k$ $0$, $k = m+1, m+2,\ldots,n$.

Proposition 2.  Let $w(z)$ be an entire transcendental solution of (25.1) of order $\rho \leq 1$.  Under the conditions enumerated above in this section $w(z)$ is a function of bounded index.

Proof.  As in the previous section we construct a g.i.  $I(r|p,a)$ the same as in Section 18 such that

$$\varlimsup_{r \to \infty} I_0(r) \leq \varlimsup_{r \to \infty} I(r|p,a) . \tag{25.3}$$

Suppose

$$\max_{k=0,1,\ldots,N} \left| \frac{g^k(z)}{k!} \right| > \left| \frac{g^j(z)}{j!} \right| \tag{25.4}$$

for each $j > N$ and every $z$ in the complex plane.

Differentiating (25.1) successively $N + 1$ time we obtain:

$$a_0(z)y^{(n+k)} + (a_0'(z) + a_1(z)y^{n+k+1}$$

$$+ \left[ \binom{k}{2} a_0''(z) + \binom{k}{1} a_1(z) + a_2(z) \right] y^{(n+k+2)} + \ldots + a_n^{(k)}(z)y = g^{(k)}(z)$$

$$k = 1,1\ldots, N+1.$$

Apply now Theorem 3.  We have

$$[a_0(\zeta)\ (\tfrac{I}{\zeta})^{(n+k)} + (a_0'(\zeta) + a_2(\zeta)(\tfrac{I}{\zeta})^{n+k+1} + \ldots + a_n^{(k)}(\zeta)]y(\zeta) = g^{(k)}(z) \tag{25.5}$$

whence in view of (25.4)

$$\max_{1 \leq k \leq N} \left| \frac{a_o(\zeta)(\frac{I}{\xi})^{n+k}+..+a_n^{(k)}(\zeta)}{k!} \right| > \left| \frac{a_o(\zeta)(\frac{I}{\xi})^{n+N+1}+..+a_n^{(N+1)}(\zeta)}{(N+1)!} \right| . \qquad (25.6)$$

Now with respect to (24.2) we have:

$$a_o(z)u^n+a_1(z)u^{n-1}+..+a_n(z) \equiv a_o(z) \prod_{k=1}^{m} (u-(1+o(1))B_k z^{\beta_k}) \prod_{k=m+1}^{n} (u-(1+o(1))B_k z^{\beta_k}). \qquad (25.7)$$

From (25.7) it follows that $\prod_{k=1}^{m}(1+o(1))B_k \bar{z}^{\beta_k} a_o(z) =$

$$= (1+o(1))A_m^o z^{\alpha_m^o} \text{ where } \alpha_m^o = \max_{1<k<n} \alpha_k^o.$$

Put now $u = \frac{I}{\zeta}$ in (25.7). Then

$$\sum_{j=0}^{n} a_j(\zeta)(\frac{I}{\xi})^{n-j}(1+o(1)) = (1+o(1))A_m^o \zeta^{\alpha_m^o}(\frac{I}{\xi})^{n-m}. \qquad (25.8)$$

By differentiation of (25.1)

$$a_o(z)y^{(n+1)}+a_1(z)y^{(n)}+..+a_n(z)y' +$$

$$+ a_o'(z)y^{(n)}+ a_1'(z)y^{(n-1)}+..+ a_n'(z)y = g'(z) . \qquad (25.9)$$

Now (25.5) for $k = 1$ gets the form

$$\{[a_o(\zeta)(1+o(1))(\frac{I}{\xi})^n + a_1(\zeta)(1+o(1))(\frac{I}{\xi})^{n-1}+..+(1+o(1))a_n(\zeta)]\frac{I}{\xi} +$$

$$+ a_o'(\zeta)(1+o(1))(\frac{I}{\xi})^n + a_j(\zeta)(1+o(1))(\frac{I}{\xi})^{n-1}+..+a_n'(\zeta)\}y(\zeta) = g'(\zeta). \qquad (25.10)$$

According to (25.8)

$$\frac{I}{\zeta} \sum_{j=0}^{n}(1+o(1))(\frac{I}{\xi})^{n-j} a_j(\zeta) = (1+o(1))A_m^o \zeta^{\alpha_m^o}(\frac{I}{\xi})^{n-m+1} . \qquad (25.11)$$

Further in the equality

$$H_o = \zeta[a_o'(\zeta)(1+o(1))(\frac{I}{\xi})^n + a_1'(\zeta)(1+o(1))(\frac{I}{\xi})^{n-1}+..+a_n'(\zeta)]$$

we have

$$\zeta a_k'(\zeta) = (1+o(1))\alpha_k A_k \zeta^{\alpha_k-1} , \quad k = 0,1,..,n,$$

so that $H_o$ is evaluated similarly to (25.8) (even if some $a_k'(\zeta)=0$; may be $a_o'(\zeta) = 0$ too) and

$$|H_o| < c|\zeta|^{\alpha_m} \left(\frac{I}{\zeta}\right)^{n-m} \tag{25.12}$$

From (25.10), (25.11) and (25.12) if follows

$$(1+o(1))A_m^o \zeta^{\alpha_m^o}\left(\frac{I}{\zeta}\right)^{n-m+1} y(\zeta) = g'(\zeta). \tag{25.13}$$

Proceeding this kind of computation it can be easily shown that asymptotically (25.5) can be rewritten as

$$(1+o(1))A_m^o \zeta^{\alpha_{mo}}\left(\frac{I}{\zeta}\right)^{n-m+k} y(\zeta) = g^{(k)}(\zeta); \qquad k = 0,1,2..N+1.$$

Now (25.6) becomes

$$\sum_{k=1}^{N} \frac{(1+o(1))\left(\frac{I}{\zeta}\right)^{n-m+k}}{k!} > \frac{(1+o(1))\left(\frac{I}{\zeta}\right)^{N+1}}{(N+1)!}$$

The last inequality is impossible since $I(r) \longrightarrow \infty$, as assumed.  So $I(r) < \infty$
$$\begin{array}{c} r \to \infty \\ r \varepsilon E' \end{array}$$

and $I_o(r) > C$.

This we have asserted in our Proposition 2.

Remark.  If deg $a_o(z) \geq$ deg $a_j(z)$, $j = 1,1...,n$, then Proposition 2 coincides with the result obtained in [10].

26.  Let us take a look at a concrete example illustrating Proposition 2 of the previous section.  Consider the equation

$$a_o(z)y^{(n)}+..+a_m(z)y^{(n-m)} - \left[\sum_{k=1}^{m} a_k(z)\right]y^{(m-1)} + z^{m-2}[y^{(m-2)} - y] =$$

$$= P_1(z)\cos\sqrt{z} + P_2(z)\sqrt{z}\sin\sqrt{z} \tag{26.1}$$

where $a_j(z)$, $P_1(z)$, $P_2(z)$ are all polynomials and such that
$q_1(z)\cos\sqrt{z} + q_2(z)\sqrt{z}\sin\sqrt{z}$ with given polynomials $q_j(z)$, $j = 1,2$ is a
solution of (26.1).  Besides $e^z$ is a solution of the homogeneous equation
corresponding to (26.1).  Thus

$$Ce^z + q_1(z)\cos\sqrt{z} + q_2(z)\sqrt{z} \text{ is a solution of (26.1).}$$

Now the function $p_1(z)\cos\sqrt{z} + p_2(z)\sqrt{z}\sin\sqrt{z}$ is an entire solution of
bounded index.  Indeed the equation

$$\left(\frac{d}{dz} - i\right)^{m_1} \left(\frac{d}{dz} + i\right)^{m_2} w = 0 \tag{26.2}$$

has a solution

$$w = p_1^*(z) \cos z + p_2^*(z) \sin z \tag{26.3}$$

with arbitrary polynomials $p_j^*(z)$, $j = 1,2$, of corresponding degrees no more than $m_j-1$, $j=1,2$ respectively.

In particular we can choose the polynomial $p_1^*(z)$ even and $p_2^*(z)$ odd. By transformation $z = \sqrt{t}$ in (26.2) one obtains an equation of type described in the remark of the previous section with a solution

$$w = p_1^{**}(z) \cos\sqrt{z} + p_2^{**}(z)\sqrt{z} \sin\sqrt{z}, \tag{26.4}$$

where $p_j^{**}(z)$ are polynomials. According to Proposition 1 of the previous section of the function (26.4) is also of bounded index.

Proposition 2 then shows that the function (26.1) is also of bounded index. Obviously, it is easy to show that each function of the type

$$\sum_{j=1}^{m_Q} P_j(z)e^{\lambda z} + \sum_{j=1}^{m_1} Q_j(z) \cos \mu_j^*\sqrt{z} + \sum_{j=1}^{m_2} R_j(z)\sqrt{z} \sin \mu_j^{**}\sqrt{z}$$

is an entire function of bounded index with arbitrary fixed numbers $\lambda_j$, $\mu_j^*$ and $\mu_q^{**}$.

## REFERENCES

1. G.H. Fricke, S.M. Shah, Entire functions satisfying a linear equation. Indag. Math. 37 (1975), 39-41.

2. G. Frank, E. Mues. Über das Wachstum des Index ganzer Funktionen. Math. Ann. 195 (1972), 114-120.

3. G.M. Golusin. Geometric theory of functions of a complex variable. Transl. Math. Monographs, AMS, Providence, R.I., 1969.

4. O. Knab. Wachstum Ordnung und Index Lösungen Linearer Differentialgleichungen mit rationalen Koeffizienten. Manuscripta Math., 18 (1976), 299-316.

5. B. Lepson, Differential equations of infinite order and entire functions of bounded index. Proc. Sympos. Pure Math. XI, AMS, Providence, R.I., (1968), 298-307.

6. B.J. Levin. Distribution of zeros of entire functions, Transl. Math. Monographs, AMS, Providence, R.I. (1964).

7. R. Nevanlinna, Remarques sur les fonctions monotones, Bull. Scien. Math. 55 (1931), 1940-1944.

8. G. Polya, G. Szegö. Problems and Theorems in Analysis, II, Springer-Verlag, N.Y. 1976.

9.   S.M. Shah, Entire functions of bounded index.  Proc. AMS 19 (1968)
     1017-1022.

10.  S.M. Shah, Entire functions of bounded index.  Lecture Notes in
     Mathematics, Complex Analysis, 599, Springer-Verlag, Kentucky (1976),
     117-141.

11.  Sh. Strelitz.  Asymptotic properties of Analytical solutions of dif-
     ferential equations, Mintis, Vilnius (USSR), 1972, Russian.

12.  G. Valiron, Fonctions Analytiques, Presses Universitaire de France,
     Paris, 1954.

Department of Mathematics
University of Haifa
Haifa, Israel

Contemporary Mathematics
Volume **25**, 1983

## ON AN EXTENSION OF THE THEOREM OF TUMURA-CLUNIE

by

Nobushige Toda

### 1. Introduction

Let $f(z)$ be a nonconstant meromorphic function in $|z| < \infty$ and $a_0$, $a_1$, ..., $b_0$, $b_1$, ..., meromorphic functions in $|z| < \infty$ of smaller growth than $f$, that is

$$T(r,a_i) = S(r,f) = T(a,b_i); \quad i = 0,1,2, \ldots , \tag{1}$$

where $S(r,f)$ is, as usual, any function satisfying

$$S(r,f) = o(T(r,f))$$

as $r \to \infty$, possibly outside a set $E$ of finite linear measure. Further, we denote by $M(f)$ the field of meromorphic functions in $|z| < \infty$ which satisfy (1).

Let

$$F \equiv a_n f^n + a_{n-1} f^{n-1} + \ldots + a_0 \quad (a_n \neq 0), \tag{2}$$

then, Clunie ([1]) proved the following

**Theorem A.** Let $f$ and $g$ be entire functions, and assume that

$$F(z) = b_0(z)\exp(g(z)).$$

Then,

$$F = a_n(f + a_{n-1}/na_n)^n.$$

This was first stated by Tumura ([6]) and proved completely by Clunie ([1]). As a generalization of this theorem, Hayman ([3], p. 69) proved a result when $F$ of (2) contains a differential polynomial in $f$ of degree at most $n-1$. Recently, Mues and Steimetz ([4]) have given the following generalization of Theorem A:

**Theorem B.** Let $f$ be a meromorphic function. Assume that $F$ given by (2) satisfies

$$\bar{N}(r,0,F) = S(r,f) \quad \text{and} \quad \bar{N}(r,f) = S(r,f). \tag{3}$$

Then

$$F = a_n(f + a_{n-1}/na_n)^n.$$

The main purpose of this paper is to give a result of Tumura-Clunie's type with a condition weaker than (3).

### 2. Lemmas

We shall give some lemmas for later use.

Lemma 1.  Let $g_0$ and $g_1$ meromorphic functions in $|z| < \infty$ which are linearly independent over C and put

$$g_0 + g_1 = \psi. \tag{4}$$

Then, we have

$$T(r,g_0) \leqq T(r,\psi) + \bar{N}(r,\psi) + \bar{N}(r,0,g_0) + \bar{N}(r,g_0) + 2\bar{N}(r,0,g_1) + 2\bar{N}(r,g_1) + S(r),$$

where

$$S(r) = \begin{cases} O(1) & \text{(when } g_0 \text{ and } g_1 \text{ are rational)}; \\ O(\log^+ T(r,g_0) + \log^+ T(r,g_1)) + O(\log r) & (r \notin E; \text{ the other cases}). \end{cases}$$

(cf. [5], Lemma 1).

Proof.  From (4) and $g_0' + g_1' = \psi'$, we have

$$g_0 = (\psi g_1'/g_1 - \psi')/(g_1'/g_1 - g_0'/g_0),$$

so that

$$\begin{aligned} T(r,g_0) &\leqq T(r, \psi g_1'/g_1 - \psi') + T(r, g_1'/g_1 - g_0'/g_0) + O(1) \\ &= m(r, \psi g_1'/g_1 - \psi') + N(r, \psi g_1'/g_1 - \psi') + m(r, g_1'/g_1 - g_0'/g_0) \\ &\quad + N(r, g_1'/g_1 - g_0'/g_0) + O(1). \end{aligned}$$

Here, we estimate each term of the right-hand side of this inequality:

$$m(r, \psi g_1'/g_1 - \psi') \leqq m(r,\psi) + m(r, \psi'/\psi) + m(r, g_1'/g_1) + O(1);$$

$$N(r, \psi g_1'/g_1 - \psi') \leqq N(r,\psi) + \bar{N}(r,\psi) + \bar{N}(r,0,g_1) + \bar{N}(r,g_1);$$

$$m(r, g_1'/g_1 - g_0'/g_0) \leqq m(r, g_1'/g_1) + m(r, g_0'/g_0) + O(1);$$

$$N(r, g_1'/g_1 - g_0'/g_0) \leqq \bar{N}(r,0,g_0) + \bar{N}(r,g_0) + \bar{N}(r,0,g_1) + \bar{N}(r,g_1).$$

Therefore, we have

$$T(r,g_0) \leqq T(r,\psi) + \bar{N}(r,\psi) + \bar{N}(r,0,g_0) + \bar{N}(r,g_0) + 2\bar{N}(r,0,g_1) + 2\bar{N}(r,g_1) + S(r),$$

where

$$S(r) = m(r,\psi'/\psi) + m(r, g_0'/g_0) + 2m(r, g_1'/g_1) + O(1)$$

$$= \begin{cases} O(1) & \text{(when } g_0 \text{ and } g_1 \text{ are rational)}; \\ O(\log^+ T(r,g_0) + \log^+ T(r,g_1)) + O(\log r) & (r \notin E, \text{ the other cases}) \end{cases}$$

(see [3]).

Lemma 2.  $T(r, \sum\limits_{j=0}^{k} a_j f^j) \leqq kT(r,f) + S(r,f)$   $(a_k \neq 0)$   (see [2], p. 46).

One can easily prove this lemma by mathematical induction.

Lemma 3.  $\bar{N}(r,F) \leqq \bar{N}(r,f) + S(r,f).$

This is easily seen from the form of F given by (2).

Lemma 4.  $1, f, \ldots, f^n$ $(n \geq 1)$ are linearly independent over $M(f)$.

This is trivial.

3.  Results

Theorem 1.  Let $f$ be meromorphic in $|z| < \infty$ and assume that

$$\limsup_{\substack{r \to \infty \\ r \notin E}} (\bar{N}(r,0,F)+2\bar{N}(r,f))/T(r,f) < 1/2. \tag{5}$$

Then

$$F = a_n(f+a_{n-1}/na_n)^n.$$

Proof.  When $n=1$, the result holds good without any condition.  So, we may suppose that $n \geq 2$.  We rewrite the right-hand side of (2) as follows:

$$F \equiv a_n(f+a_{n-1}/na_n)^n + \sum_{j=0}^{n-2} b_j f^j,$$

where $b_j (j=0, \ldots, n-2)$ are rational functions in $a_j, a_{n-1}, a_n$.

Suppose that at least one of $b_j$ is not equal to zero identically and let $\nu$ be the maximum number of $j$ such that $b_j \neq 0$.  Put

$$\psi \equiv \sum_{j=0}^{\nu} b_j f^j \quad (0 \leq \nu \leq n-2, \ b_\nu \neq 0),$$

then by Lemma 4, $\psi \neq 0$.  Let

$$g_0 \equiv -a_n(f+a_{n-1}/na_n)^n \text{ and } g_1 \equiv F,$$

then we can easily prove that $g_0$ and $g_1$ are linearly independent over C by Lemma 4 and

$$g_0 + g_1 = \psi.$$

By Lemma 1,

$$T(r,g_0) \leq T(r,\psi)+\bar{N}(r,\psi)+\bar{N}(r,0,g_0)+\bar{N}(r,g_0)+2\bar{N}(r,0,g_1)+2\bar{N}(r,g_1)+S(r),$$

Here, we use the following estimates.

$$T(r,g_0) = nT(r,f)+S(r,f);$$

$$T(r,\psi) \leq \nu T(r,f)+S(r,f) \quad \text{(by Lemma 2)};$$

$$\bar{N}(r,0,g_0) \leq \bar{N}(r,0,f+a_{n-1}/na_n)+S(r,f) \leq T(r,f)+S(r,f);$$

$$\bar{N}(r,g_0) \leq \bar{N}(r,f)+S(r,f);$$

$$\bar{N}(r,0,g_1) = \bar{N}(r,0,F);$$

$$\bar{N}(r,g_1) \leq \bar{N}(r,f)+S(r,f) \quad \text{(by Lemma 3)};$$

$\bar{N}(r,\psi) \leqq \bar{N}(r,f)+S(r,f)$    (by Lemma 3).

Then, we obtain the inequality:

$$T(r,f) \leqq 4\bar{N}(r,f)+2\bar{N}(r,0,F)+S(r,f),$$

and we get

$$1 \leqq \lim_{\substack{r \to \infty \\ r \notin E}} \sup (2\bar{N}(r,0,F)+4\bar{N}(r,f))/T(r,f) < 1,$$

which is a contradiction.  This shows that all $b_j$ must be equal to zero

identically, that is,

$$F = a_n(f+a_{n-1}/na_n)^n. \tag{6}$$

Corollary.  If   $2n+3 < 2n\delta(0,F)+4\delta(\infty,f)$,

then

$$F = a_n(f+a_{n-1}/na_n)^n.$$

In fact, it is easily seen that (6) implies (5).

Theorem 2.  Let $f$ be meromorphic in $|z| < \infty$ of order $\rho(0<\rho\leqq\infty)$.  Suppose, instead of (1), that the orders of $a_j$ are less than $\rho$ and assume that the orders of $\bar{N}(r,f)$ and $\bar{N}(r,0,F)$ are less than $\rho$.  Then

$$F = a_n(f+a_{n-1}/na_n)^n.$$

By applying Lemma 1 to this case and in a manner similar to the treatment of Theorem 1, we obtain this theorem.

## REFERENCES

[1]  J. Clunie, On integral and meromorphic functions, J. London Math. Soc., 37, 17-27 (1962).

[2]  A.A. Gol'dberg and I.V. Ostrowskii, The distribution of values of meromorphic functions, Nauka, Moskow (1970).

[3]  W.K. Hayman, Meromorphic functions, Oxford at the Clarendon Press (1964).

[4]  E. Mues and N. Steimetz, The theorem of Tumura-Clunie for meromorphic functions, J. London Math. Soc., 23, 113-112 (1981).

[5]  N. Toda, On the conjecture of Gackstatter and Laine concerning the differential equation $(w')^n = \sum_{j=0}^{m} a_j w^j$, Kodai Math. J. 6 (1983).

[6]  Y. Tumura, On the extensions of Borel's theorem and Saxer-Csillag's theorem, Proc. Phys. -Math. Soc. Japan, 19, 29-35 (1937).

DEPARTMENT OF MATHEMATICS
NAGOYA INSTITUTE OF TECHNOLOGY
GOKISO, SHOWA-KU, NAGOYA 466
JAPAN

Contemporary Mathematics
Volume **25**, 1983

## ON A PROBLEM OF ANDERSON AND RUBEL

Sakari Toppila

### 1. Introduction and results

We shall consider the following problem of Anderson and Rubel (1, Problem 2.53):

Problem. For entire or, more generally, meromorphic functions f and g let "f $\leq$ g" mean that, for any sequence $z_n$ for which $|f(z_n)| \to \infty$, then $|g(z_n)| \to \infty$. For entire functions it can be proved that, if f' $<$ f, then f is of exponential type; what can be said if f" $\leq$ f? Does f'$\leq$ f imply that f is normal (that is, that

$$|f'(z)|(1+|f(z)|^2)^{-1}$$

is bounded)? In the above ordering does there exist f $\wedge$ g and f $\vee$ g for any two entire functions f and g? That is, given f and g, does there exist an h such that h $\leq$ f and h $\leq$ g and so that if k $\leq$ f and k $\leq$ g, then k $\leq$ h? Similarly for f $\vee$ g. Finally, if f is meromorphic and f' $\leq$ f, does this imply a growth restriction on f, e.g., is the order of f at most two?

We shall give a positive answer for two questions of this problem.

Theorem 1. Let f be a meromorphic function in the plane such that f' $\leq$ f. Then

$$|f'(z)|(1+|f(z)|^2)^{-1}$$

is bounded and

$$T(r,f) = O(r^2) \quad (r \longrightarrow \infty) \tag{1.1}$$

For entire functions, we shall prove the following results.

Theorem 2. There exist entire functions f and g such that f $\wedge$ g does not exist.

Theorem 3. There exist entire functions f and g such that if h is any entire function such that f $\leq$ h and g $\leq$ h then there exists an entire function k such that f $\leq$ k, g $\leq$ k and that k does not satisfy the condition h $\leq$ k.

### 2. Proof of Theorem 1

Lemma 1. Suppose that f is meromorphic and $|f(z)| \leq 1$ on $|z| \leq r$, and that there exists $z_0$ lying on $|z| = r$ such that $|f(z_0)| = 1$. Then

$$|f'(0)| \leq 4|f'(z_0)|. \tag{2.1}$$

Proof. It does not mean any restriction to assume that r = 1, because in other case we may consider the function f(rz) in stead of f(z). Furthermore, we may assume that $|f(0)| < 1$, because if $|f(0)| = 1$ then f is a constant function.

We set

$$h(z) = \frac{f(z) - f(0)}{1 - \overline{f(0)}f(z)} .$$

It follows from Schwarz' lemma that

$$|h'(0)| \leq 1 \tag{2.2}$$

and

$$|h(z)| \leq |z| \text{ for } |z| \leq 1. \tag{2.3}$$

Let $0 < t < 1$. We deduce from (2.3) that

$$1 - t = |z_0 - tz_0| \geq |h(z_0 - tz_0)|$$

$$= |h(z_0) - (h'(z_0) + o(1))tz_0|$$

$$\geq 1 - (|h'(z_0)| + o(1))t \quad (t \longrightarrow 0)$$

which implies that

$$|h'(z_0)| \geq 1. \tag{2.4}$$

Since

$$h'(z) = \frac{f'(z)(1 - |f(0)|^2)}{(1 - \overline{f(0)}f(z))^2}$$

we get from (2.2) and (2.4)

$$\left| \frac{f'(0)}{f'(z_0)} \right| = \left| \frac{h'(0)(1 - |f(0)|^2)^2}{h'(z_0)(1 - \overline{f(0)}f(z_0))^2} \right|$$

$$\leq \left( \frac{1 - |f(0)|^2}{1 - |f(0)|} \right)^2 = (1 + |f(0)|)^2 \leq 4,$$

which proves (2.1). Lemma 1 is proved.

Let $f$ be meromorphic in the plane. We write

$$\psi(f(z)) = |f'(z)| (1 + |f(z)|^2)^{-1}$$

and

$$E(f) = \{z : |f(z)| = 1\} .$$

From Lemma 1 we get the following

Corollary 1. Let $f$ be a non-constant meromorphic function in the finite complex plane C. Then

$$\sup \{\psi(f(z)) : z \in C\} \leq 4\sup \{|f'(z)| : z \in E(f)\} . \tag{2.5}$$

Proof.  Let $z_1 \in C$ be given.  If $|f(z_1)| < 1$ we choose $r > 0$ such that $|f(z)| < 1$ in $|z - z_1| < r$ and that $|f(z_0)| = 1$ for some $z_0$ lying on $|z - z_1| = r$.  From Lemma 1 we get

$$\psi(f(z_1)) \leq |f'(z_1)| \leq 4|f'(z_0)|$$

where $z_0 \in E(f)$.  Similarly, if $|f(z_1)| > 1$, considering $1/f$ instead of $f$, we deduce that there exists $z_0 \in E(f)$ such that

$$\psi(f(z_1)) \leq |f'(z)/f^2(z)| \leq 4|f'(z_0)/f^2(z_0)|$$

$$= 4|f'(z_0)|.$$

From these estimates we get (2.5).  Corollary 1 is proved.

Proof of Theorem 1.  Let $f$ be a non-constant meromorphic function such that $f' \leq f$.  Then

$$\sup\{|f'(z)| : z \in E(f)\} = M < \infty$$

and we deduce from Corollary 1 that

$$\psi(f(z)) \leq 4M \tag{2.6}$$

for all $z \in C$.  This proves the first assertion of Theorem 1.

Clunie and Hayman [2] have proved that if

$$|z|\, \psi(f(z)) = O(|z|^d) \quad (z \longrightarrow \infty)$$

for some $d > 0$ then

$$T(r,f) = O(r^{2d}) \quad (r \longrightarrow \infty),$$

so that, we get (1.1) from (2.6).  Theorem 1 is proved.

### 3.  Some lemmas

Lemma 2.  Let $c_n$ be an increasing sequence such that $c_0 > 9$ and that

$$c_n \geq \operatorname{expexp}(c_{n-1}) \tag{3.1}$$

for $n \geq 1$, and let $f$ be a transcendental entire function such that

$$|f(z)| > 2 + o(1) \tag{3.2}$$

as $z \longrightarrow \infty$ outside the union of the discs $|z - c_n| < c_n/9$.  Then

$$T(r,f) = o((\log r)^2 \log\log r) \quad (r \longrightarrow \infty) \tag{3.3}$$

and

$$n(r,0,f) = o((\log r)\log\log r) \quad (r \longrightarrow \infty). \tag{3.4}$$

Proof.   We use the usual notations of the Nevanlinna theory and we write

$$L(r,f) = \min\{|f(z)| : |z| = r\}.$$

It follows from Schottky's theorem and (3.2) that there exists an absolute constant $A_1 > 0$ such that

$$\log M(4c_n,f) \leq A_1 \log L(c_n/2,f)$$

for all large values of n.  This implies together with the first main theorem of the Nevanlinna theory and (3.2) that

$$N(4c_n,0,f) = T(4c_n,f) + O(1) \tag{3.5}$$

$$\leq \log M(4c_n,f) + O(1)$$

$$\leq A_1 \log L(c_n/2,f) + O(1)$$

$$\leq A_1 T(c_n/2,f) + O(1)$$

$$= A_1 N(c_n/2,0,f) + O(1) \quad (n \longrightarrow \infty).$$

From (3.5) and (3.2) we deduce that

$$n(2c_p,0,f)\log 2 = \int_{2c_p}^{4c_p} n(t,0,f)t^{-1}dt \leq N(4c_p,0,f)$$

$$= A_1 N(c_p/2,0,f) + O(1) \leq A_1 n(c_p/2,0,f)\log c_p + O(1)$$

$$= A_1 n(2c_{p-1},0,f)\log c_p + O(1) \quad (P \longrightarrow \infty).$$

This implies together with (3.1) that there exists $p_0 > 0$ such that

$$\frac{n(2c_p,0,f)}{(\log c_p)\,\log\log c_p} \leq \frac{(A_1 + o(1))n(2c_{p-1},0,f)}{c_{p-1}\log 2}$$

$$\leq \frac{n(2c_{p-1},0,f)}{2(\log c_{p-1})\log\log c_{p-1}}$$

for $p \geq p_0$.   This implies that

$$n(2c_p,0,f) = o((\log c_p)\log\log c_p) \quad (p \longrightarrow \infty)$$

which together with the fact that

$$n(c_{p+1}/2,0,f) = n(2c_p,0,f)$$

for all large p, proves (3.4).

From (3.4) we get

$$N(r,0,f) = O(n(r,0,f)\log r)$$

$$= o((\log r)^2 \log\log r) \quad (r \longrightarrow \infty)$$

and since

$$T(r,f) = (1 + o(1))N(r,0,f)$$

as $r \longrightarrow \infty$ outside the intervals $|r - c_n| \leq c_n/9$, we deduce that (3.3) holds.

Lemma 2 is proved.

Lemma 3. Let $r_p$ be an increasing sequence such that $r_0 = 100$ and that

$$r_p \geq \text{expexpexp}(r_{p-1}) \tag{3.6}$$

for $p \geq 1$, and let

$$f(z) = \prod_{p=1}^{\infty} (1-z/r_p).$$

Then

$$|f(z)| \longrightarrow \infty \tag{3.7}$$

as $z \longrightarrow \infty$ outside the union of the discs $|z - r_p| < 1/r_p$,

$$\log|f(z)| = N(r_p,0,f) + \log|(z - r_p)/r_p| + o(1) \tag{3.8}$$

in $|z - r_p| < 1 \quad (p \longrightarrow \infty)$, and

$$N(r_p,0,f) = (1 + o(1))p \log r_p \quad (p \longrightarrow \infty). \tag{3.9}$$

Proof. Let $|z - r_p| < 1$, $p > 2$. It follows from (3.6) that

$$\left| \log|f(z)| - N(r_p,0,f) - \log|(z - r_p)/r_p| \right|$$

$$\leq \left| \sum_{k=1}^{p-1} \log|1 - z/r_k| - \sum_{k=1}^{p-1} \log(r_p/r_k) \right| + o(1)$$

$$\leq \left| \sum_{k=1}^{p-1} \log|r_k/r_p - 1 - (z - r_p)/r_p| \right| + o(1)$$

$$= o((pr_{p-1})/r_p) + o(1) = o(1) \quad (p \longrightarrow \infty)$$

which proves (3.8).

From (3.6) we get

$$N(r_p,0,f) = \sum_{k=1}^{p-1} \log(r_p/r_k)$$

$$= (1 + o(1))p \log r_p \quad (p \longrightarrow \infty)$$

$$\lim_{p \to \infty} L(r_p, f) = \infty,$$

we get (3.7) from the minimum principle.  Lemma 3 is proved.

For any real x, we denote by x that integer which satisfies
$x - 1 < [x] \le x$.

Lemma 4.  Let $r_0 = 100$ and

$$r_p = \text{expexpexp}(pr_{p-1}) \tag{3.10}$$

for $p \ge 1$.  Let $d_p$ be a sequence of positive real numbers such that $d_p < 1$ for all p and that

$$\log(r_p/d_p) = (1/2 + o(1))p \log r_p \quad (p \longrightarrow \infty). \tag{3.11}$$

We choose

$$s_p = [r_p^{1/2}(\log(r_p/d_p))^{-1}]$$

and set

$$H(z) = (\exp(\sqrt{z}) + \exp(-\sqrt{z})) \prod_{p=1}^{\infty} (1 - z/r_p)^{s_p}.$$

Then

$$|H(z)| \longrightarrow \infty \tag{3.12}$$

as $z \to \infty$ outside the union of the angle

$$|\pi - \arg z| < \pi/4 \text{ and the discs } |z - r_p| < 2d_p, \text{ and}$$

$$|H(z)| \longrightarrow 0 \tag{3.13}$$

as $z \to \infty$ through the union of the discs $|z - r_p| < d_p/2$.

Proof.  Let $|z - r_p| \le 2d_p$.  It follows from (3.10) and (3.11) that

$$\log|H(z)| = r_p^{1/2} + s_p \log|(z - r_p)/r_p| \tag{3.14}$$

$$+ O(s_{p-1} \log r_p) \quad (p \longrightarrow \infty).$$

This implies together with (3.10) and (3.11) that if $|z - r_p| \le d_p/2$ then

$$\log|H(z)| \le r_p^{1/2} - (r_p^{1/2}(\log(r_p/d_p))^{-1} - 1)\log(2r_p d_p^{-1})$$

$$+ O(s_{p-1} \log r_p)$$

$$\le - r_p^{1/2}(\log(r_p/d_p))^{-1}\log 2$$

$$+ O(\log(r_p/d_p) + s_{p-1} \log r_p)$$

$$\le - r_p^{1/4} + o(1) \quad (p \longrightarrow \infty)$$

which proves (3.13).

Let $|z - r_p| = 2d_p$. From (3.14), (3.10) and (3.11) we get

$$\log|H(z)| \geq r_p^{1/2} - r_p^{1/2}(\log(r_p/d_p))^{-1}\log(r_p(2d_p)^{-1}) \tag{3.15}$$

$$+ O(s_{p-1}\log r_p)$$

$$\geq r_p^{1/4} + o(1)) \quad (p \longrightarrow \infty).$$

Using the minimum principle, we get (3.12) from (3.15) and the facts that

$$\lim_{r \to \infty}|H(r\exp((3\pi i)/4))| = \infty,$$

$$\lim_{r \to \infty}|H(r\exp((5\pi i)/4))| = \infty,$$

and that $|H(z)| \longrightarrow \infty$ as $z \longrightarrow \infty$ through the arcs

$$\{z : |z| = r_p^{1/4}, \ |\pi - \arg z| \geq \pi/4 \}.$$

Lemma 4 is proved.

Lemma 5. Let $r_p$ be as in Lemma 4. We choose

$$f(z) = \prod_{p=1}^{\infty} (1 - z/r_{2p-1})$$

and

$$g(z) = \prod_{p=1}^{\infty}(1 - z/r_{2p}).$$

Given any sequence $z_p$ such that $0 < |z_p - r_p| \leq 1$ for all $p$ and that

$$\lim_{p \to \infty} \min\{ |f(z_p)|, |g(z_p)| \} = \infty, \tag{3.16}$$

there exists an entire function $H$ such that $H \leq f$, $H \leq g$ and that

$$|H(z)| \longrightarrow \infty \tag{3.17}$$

as $z \longrightarrow \infty$ outside the union of the angle $|\pi - \arg z| < \pi/4$ and the discs $|z - r_p| < |z_p - r_p|$.

Proof. We set

$$M_p = \min\{ p, |f(z_p)|, |g(z_p)| \}.$$

Then $M_p \longrightarrow \infty$ as $p \longrightarrow \infty$. It follows from Lemma 3 that for large values pf $p$, say for $p > p_0$, there exists $d_p$, $0 < d_p < 1/4$, such that

$$|f(r_p + 4d_p)| = M_p \text{ if } p \text{ is odd, and} \tag{3.18}$$

228                          S. TOPPILA                                     (3.19)

$$g(r_p + 4d_p) = M_p \text{ if } p \text{ is even.}$$

Let $d_p = 1/8$ for $p = 1, \ldots, p_0$.

It follows from (3.18) and Lemma 3 that

$$\log(r_{2p-1}/(4d_{2p-1})) = N(r_{2p-1}, 0, f)$$
$$- \log M_{2p-1} + o(1)$$
$$= (1/2 + o(1))(2p - 1)\log r_{2p-1} \quad (p \longrightarrow \infty),$$

and since we get a similar estimate for g, we deduce that

$$\log(r_p/d_p) = (1/2 + o(1))p \log r_p \quad (p \longrightarrow \infty).$$

Let H be as in Lemma 4 corresponding to this choice of $d_p$. Then H is bounded on the union of the discs $|z - r_p| < d_p/2$. Let p be odd and $d_p/2 \leq |z - r_p| < 1$. It follows from (3.18) and Lemma 3 that

$$\log|f(z)| = \log|f(z)/f(r_p + 4d_p)| + \log M_p$$
$$\geq \log(1/8) + \log M_p + o(1) \quad (P \longrightarrow \infty)$$

which together with Lemma 3 implies that $|f(z)| \longrightarrow \infty$ as $z \longrightarrow \infty$ outside the discs $|z - r_p| \leq d_p/2$. Similarly, we see that $|g(z)| \to \infty$ as $z \longrightarrow \infty$ outside the discs $|z - r_p| \leq d_p/2$, and we deduce that $H \leq f$ and $H \leq g$.

From the choice of $M_p$, Lemma 3 and (3.18) we deduce that if p is odd then

$$0 \leq \log|f(z_p)/f(r_p + 4d_p)| = \log|(z_p - r_p)/(4d_p)| + o(1)$$
$$= \log|(z_p - r_p)/(2d_p)| - \log 2 = o(1) \quad (p \longrightarrow \infty)$$

which implies that
$$|z_p - r_p| > 2d_p \qquad\qquad\qquad (3.20)$$

for all large values of p. From (3.20) and Lemma 4 we deduce that (3.17) holds. Lemma 5 is proved.


## 4.  Proof of Theorem 2

Let f and g be as in Lemma 5. Contrary to the assertion of Theorem 2, let us suppose that there exists an entire function $h = f \wedge g$. Since $e^{-z} \leq f$ and $e^{-z} \leq g$ we deduce that $e^{-z} \leq h$, which implies that $|h(z)| \longrightarrow \infty$ as $z \longrightarrow \infty$ on the angle $|\arg z - \pi| \leq \pi/4$.

If follows from Lemma 3 that

$$\lim_{p\to\infty} \min\{|f(r_p +/r_p)|, |g(r_p + 1/r_p)|\} = \qquad (4.1)$$

which together with Lemma 5 implies that there exists an entire function H such that $H \le f$, $H \le g$ and that $|H(z)| \longrightarrow \infty$ as $z \longrightarrow \infty$ outside the union of the angle $|\arg z - \pi| < \pi/4$ and the discs $|z - r_p| < 1/r_p$. Since $H \le h$ and $e^{-z} \le h$, we deduce that

$$|h(z)| \longrightarrow \infty \qquad (4.2)$$

as $z \longrightarrow \infty$ outside the discs $|z - r_p| < 1/r_p$.

From Lemma 2 and (4.2) it follows that

$$h(z) = P(z) \prod_{p=p_1}^{\infty} \prod_{k=1}^{k_p} (1 - z/a_{p,k})$$

where $P(z)$ is a polynomial and $|a_{p,k} - r_p| < 1/r_p$ for $k = 1, \ldots, k_p$. Since $h \le f$ and $h \le g$, we deduce that $k_p \ge 1$ for all large values of $p$, and we may assume that $P(z)$ is chosen such that $k_p \ge 1$ for $p \ge p_1$. This together with (4.2) implies that for any large $p$, there exists $t_p < 1/r_p$ such that

$$|h(z)| > 1 \text{ for } t_p < |z - r_p| < 1 \qquad (4.4)$$

and that

$$|h(b_p)| = 1 \qquad (4.5)$$

for some $b_p$ lying on $|z - r_p| = t_p$.

Let us suppose that

$$\limsup_{p\to\infty} \min\{|f(b_p)|, |g(b_p)|\} = \infty . \qquad (4.6)$$

We choose a subsequence $b_{p_s}$ such that

$$\lim_{p\to\infty} \min\{|f(b_{p_s})|, |g(b_{p_s})|\} = \infty \qquad (4.7)$$

and a sequence $z_p$ such that $z_p = b_{p_s}$ if $p = p_s$ for some s and that $z_p = r_p + 1/r_p$ for the other values of $p$. From Lemma 5, (4.1) and (4.7) it follows that there exists an entire function G such that $G \le f$, $G \le g$ and that

$$\lim_{p\to\infty} |G(z_p)| = \infty.$$

Since $G \le h$, we get

$$\limsup_{p\to\infty} |h(b_p)| = \infty .$$

This is a contradiction with (4.5) and we deduce that (4.6) does not hold. This implies that there exists $M < \infty$ such that

$$\min\{|f(b_p)|, |g(b_p)|\} < M \tag{4.8}$$

for all large p.

From Lemma 3 and (4.8) we get

$$\min\{\log|f(r_p + 9t_p)|, \log|g(r_p + 9t_p)|\}$$

$$\leq \log((9t_p)/t_p) + \log M + o(1) = O(1) \qquad (P \to \infty)$$

and since $h \leq f$ and $h \leq g$ we deduce that

$$|h(r_p + 9t_p)| = O(1) \qquad (P \to \infty).$$

This together with Lemma 2 implies that

$$O(1) \geq \log h(r_p + 9t_p)/h(b_p)$$

$$\geq k_p \log((8t_p)/(2t_p)) + O(n(2r_{p-1},0,h)r_p^{-1})$$

$$\geq k_p \log 4 + o(1) \qquad (P \to \infty)$$

and we deduce that there exists $k_0 < \infty$ such that

$$k_p \leq k_0 \text{ for all } p \geq p_1. \tag{4.9}$$

It follows from Lemma 3 that there exists a sequence $d_p$, $0 < d_p < 1$, such that

$$\log(r_p/d_p) = (1/2 + o(1))p \log r_p \qquad (p \to \infty) \tag{4.10}$$

and that the function

$$\min\{|f(z)|, |g(z)|\}$$

is bounded on the union of the discs $|z - r_p| \leq d_p$. Then h is bounded on these discs, too. From (4.9) we deduce that for any large p, there exists $w_p$ such that $r_p \leq w_p \leq r_p + d_p$ and that

$$|w_p - a_{p,k}| \geq d_p/(2k_0 + 2)$$

for $k = 1, \ldots, k_p$. This together with (4.9) implies that

$$O(1) \geq \log|h(w_p)|$$

$$\geq (1 + o(1))(\log r_p) \sum_{s=p_1}^{p-1} k_s$$

$$\qquad - k_p \log((2r_p(2rk_0 + 2))/d_p)$$

$$\geq (1 + o(1))(\log r_p) \sum_{s=P_1}^{p-1} k_s - k_p \log(r_p/d_p)$$

as $p \longrightarrow \infty$ , which together with (4.10) implies that

$$k_p \geq (2 + o(1))p^{-1} \sum_{s=P_1}^{p-1} k_s \quad (p \longrightarrow \infty). \tag{4.11}$$

Let

$$B = \liminf_{p \to \infty} k_p .$$

Then $B \geq 1$ and we deduce from (4.11) and (4.9) that

$$k_0 \geq B \geq 2B > 1.$$

This is a contradiction which shows that the function $h = f \wedge g$ does not exist.   Theorem 2 is proved.

5.   Proof of Theorem 3

Let $r_0 = 100$ and

$$r_p = \text{expexpexp}(r_{p-1}) \tag{5.1}$$

for $p \geq 1$.  We set $g(z) = e^{-z}$ and

$$f(z) = (\exp(\sqrt{z}) + \exp(-\sqrt{z})) \prod_{p=1}^{\infty} (1 - z/r_p)^{s_p}$$

where

$$s_p = 1 + [r_p^{1/2}(2\log r_p)^{-1}] . \tag{5.2}$$

Let $d_p = (1 - r_p^{-1/4})/r_p$ and let $|z - r_p| = d_p$. We have

$$\log|f(z)| \leq r_p^{1/2} + \sum_{k=1}^{p-1} s_k \log(r_p/r_k)$$

$$+ s_p \log(d_p/r_p) + o(1)$$

$$\leq r_p^{1/2} + (\log r_p) \sum_{k=1}^{p-1} s_k + o(1)$$

$$- r_p^{1/2}(2\log r_p)^{-1}(2 \log r_p + (2r_p)^{-1/4})$$

$$\leq 0((\log r_p)^2) - r_p^{1/5} + o(1)$$

$$\leq o(1) \quad (p \longrightarrow \infty)$$

which implies that f is bounded on the union of the discs $\left|z - r_p\right| \leq d_p$.

Let $\left|z - r_p\right| = 1/r_p$.  We have

$$\log\left|f(z)\right| \geq r_p^{1/2} + (s_{p-1}/2)\log r_p - 2s_p\log r_p + o(1)$$

$$\geq (s_{p-1}/2 - 2)\log r_p + o(1) \quad (p \longrightarrow \infty)$$

which implies that

$$\lim_{p \longrightarrow \infty} \min\{\ \left|f(z)\right|\ :\ \left|z - r_p\right| = 1/r_p\} = \infty \ . \tag{5.3}$$

From (5.3) we deduce that $f(z) \longrightarrow \infty$ as $z \longrightarrow \infty$ outside the union of the angle $\arg\left|z - \pi\right| \leq \pi/4$ and the discs $\left|z - r_p\right| < 1/r_p$.

Let us suppose now that there exists an entire function h such that $f \leq h$, $g \leq h$ and that $h \leq k$ for any entire function k satisfying the conditions $f \leq k$ and $g \leq k$.

Since $f \leq h$ and $g \leq h$, we deduce that

$$h(z) \longrightarrow \infty \tag{5.4}$$

as $z \longrightarrow \infty$ outside the discs $\left|z - r_p\right| < 1/r_p$.

Let $a_p$ be any sequence such that

$$\left|a_p - r_p\right| \leq d_p - r_p^{-5/4} = r_p^{-1}(1 - 2r_p^{-1/4}) \tag{5.5}$$

for all p.  We set

$$k(z) = \prod_{p=1}^{\infty} (1 - z/a_p).$$

Then $k(z) \longrightarrow \infty$ as $z \longrightarrow \infty$ outside the union of the discs $\left|z - a_p\right| < r_p^{-2}$ and therefore $k(z) \longrightarrow \infty$ as $z \longrightarrow \infty$ outside the discs $\left|z - r_p\right| < d_p$.  We deduce that $f \leq k$ and $g \leq k$ and we get $h \leq k$.  This implies that

$$\limsup_{p \to \infty} \left|h(a_p)\right| < \infty$$

for any sequence $a_p$ satisfying (5.5) and we deduce that h is bounded on the union of the discs

$$T_p = \{\ z\ :\ \left|z - r_p\right| \leq r_p^{-1}(1 - 2r_p^{-1/4})\}\ .$$

This together with (5.4) and Lemma 2 implies that h is an entire function which satisfies

$$T(r,h) = O((\log r)^3) \quad (r \longrightarrow \infty),$$

which implies that

$$\log M(r,h) = O((\log r)^3) \quad (r \longrightarrow \infty). \tag{5.6}$$

Since h is bounded on the union of the discs $T_p$, there exists $K < \infty$ such that $|h(z)| \leq K$ for all z lying on the union of $T_p$. On the boundary of the annulus

$$D_p = \{ z : r_p^{-1}(1 - 2r_p^{-1/4}) < |z - r_p| < r_p/4 \}$$

we have

$$\log|h(z)| \leq \log K + (\log M(2r_p,f))\log \frac{r_p|z - r_p|}{1 - 2r_p^{-1/4}}, \tag{5.7}$$

and from the maximum principle we deduce that (5.7) holds in $D_p$. From (5.6) and (5.7) we get

$$\log|h(r_p + 1/r_p)|$$
$$\leq \log K + (\log M(2r_p,f))\log(1 - 2r_p^{-1/4-1})^{-1}$$
$$\leq \log K + O(r_p^{-1/4}(\log r_p)^3)$$
$$\leq \log K + o(1) \quad (p \longrightarrow \infty).$$

This is a contradiction with (5.4), and we deduce that the functions f and g satisfy of Theorem 3. Theorem 3 is proved.

## References


1.  J.M. Anderson, K.F. Barth and D.A. Brannan, "Research problems in complex analysis", Bull. London Math. Soc. 9 (1977), 129 - 162.

2.  J. Clunie and W. K. Hayman, "The spherical derivative of integral and meromorphic functions", Comment. Math. Helv. 40 (1966), 117 - 148.

3.  Ch. Pommerenke, "Normal functions", Proc. of the NRL Conference on Classical Function Theory (1970), 77 - 93.



Department of Mathematics
University of Helsinki
SF - 00100 Helsinki 10
Finland

Contemporary Mathematics
Volume **25**, 1983

## ON ANALYTIC COVERING SPACES OF $C^m$

Chia-Chi Tung

It is well-known that every Stein space admits a biholomorphic embedding into some $C^n$. On the other hand, it is also known that not every m-dimensional Stein space admits a proper, holomorphic map into $C^m$. For instance, every bounded domain in $C^m$ ($m > 0$) is of this type ([6]). One is thus lead to the natural question: which (Stein) complex space can be realized as an analytic cover of $C^m$? A characterization of such spaces, in the irreducible case, shall be given in terms of the growth of the space relative to suitable exhaustion functions.

A complex space X is said to be of <u>type E</u> if every compacy analytic subset of X is finite. For example, if X carries a $C^2$-strongly plurisubharmonic function or if X is holomorphically separable, then X is of type E ([3,p. 464] [4, pg. 121]). If $\Psi$ is an exhaustion function of X, set $X[r] = \{\Psi \leq r\}$, $X(r) = \{\Psi < r\}$, etc. Also define $\Theta = dd^c\Psi$, where $d^c = (i/4\pi)(\bar{\partial} - \partial)$. The main result obtained is the following:

<u>Theorem</u> I. Let X be an irreducible complex space of dimension m $\quad$ 0. Then X is an analytic cover of $C^m$ if (i) X is of type E; (ii) X admits a proper, holomorphic map f into some $C^n$ and there exists a continuous function $Q$: $R(0,\infty) \longrightarrow R$ such that

(a)  $Q \in C^1(R(0,\infty))$ and $Q(r) \geq c_0 = $ const. $> 0$ for large r;

(b)  relative to the exhaustion $\Psi = \| f \|^2$,

$$\int_{X(C,Y)} (\Psi * Q)\Omega^m + \int_{c_1}^{Y} \left( \int_{X(t)} \Theta^m \right) \frac{Q'(t)}{t^m} \, dt = O(Q(r)) \tag{0.1}$$

as $r \longrightarrow \infty$, for some positive constants $c, c_1$, where $\Omega = dd^c \log \Psi$.

As a tool in proving the above result, a Cauchy type integral formula is first derived for a space with a centrally g-convex exhaustion function. Applied locally to a complex space, this formula also yields a characterization of its simple points.

## 1. A Cauchy type integral formula

In what follows let $X$ be a reduced, pure $m$-dimensional complex space which is countable at infinity. Let $g: R(0, \infty) \longrightarrow R$ be increasing of class $C^1$. Assume $\Psi : X \longrightarrow R(-\infty, b)$ (where $0 < b \leq \infty$) is a $C^\infty$-exhaustion function such that $dd^c\Psi \geq \Psi * g' \, d\Psi \wedge d^c\Psi$ in $X_* = X - X[0]$. For $q \in Z[1,m]$ define

$$\Omega = \Psi * u \, [\Theta - \Psi * g' \, d\Psi \wedge d^c\Psi] \tag{1.1}$$

$$B_q = \frac{-i}{2\pi} \; \Psi * u \; \Omega^{q-1} \wedge \partial\Psi \tag{1.2}$$

$$\Sigma_q = \Psi * u \, \Omega^{q-1} \wedge d^c\Psi \tag{1.3}$$

on $X_*$, where $u = e^{-g}$. It follows that

$$\Omega^q = (\Psi * u)^q \, \Theta^{q-1} \wedge [\Theta - q\Psi * g' \, d\Psi \wedge d^c\Psi]. \tag{1.4}$$

Let $A_\Psi$ be the set of all positive $\Psi$-admissible values ([11, p. 538]). If $r \in A_\Psi$, let $dX(r)$ be the maximal boundary manifold of $X(r)_{reg}$ ([11, p. 534]) and $j_r: dX(r) \longrightarrow X_{reg}$ the inclusion map. Then $(1.3)$-$(1.4)$ and $[8, 3.2]$ imply that

$$j_r^*(\Sigma_m) \quad u(r)^m \, (\Theta^{m-1} \wedge d^c\Psi) \geq 0. \tag{1.5}$$

For continuous functions $h$ and $k : X_* \longrightarrow C$ with $k \in C^1(X_*)$, define

$$B_{h,k}(r) = \int_{dX(r)} h \, \Omega^{m-1} \wedge d^c k \qquad (r \in A_\Psi).$$

(For the existence of the integral see [10, 7.1.8]). Also set

$$M_h(r) = B_{h,w}(r) \qquad \text{with } w = \Psi * U, \tag{1.6}$$

where $U(r) = \| u \|_{L^1[1,r]}$, and

$$P_k(r) = B_{1,k}(r) \tag{1.7}$$

for $r \in A_\Psi$. Observe that by $(1.4)$-$(1.6)$,

$$M_h'(r) = \int_{dX(r)} h \, \Sigma_m. \tag{1.8}$$

Lemma 1.1. Assume $\Psi$ is real analytic. Let $h \in C^1(X_*)$. Then for all $r, s$ in $A_\Psi$ with $s < r$,

$$M_h(r) - M_h(s) - \int_{X(s,r)} h\,\Omega^m = \int_s^r P_h(t)\,u(t)\,dt \qquad (1.9)$$

$$= \int_{X(s,r)} dh \wedge \Sigma\,m .$$

Proof.  It will be shown that the form $\Omega^{m-1} \wedge d^c h$ admits a decomposition

$$\Omega^{m-1} \wedge d^c h = F_1 + F_2 - F_3 \qquad (1.10)$$

a.e. on $(X_*)_{reg}$, with the properties:

$$d\Psi \wedge F_1 = 0, \qquad j_r^*(F_1) = 0,$$

$$j_r^*(F_n) \geq 0 \text{ for } n = 2,3, \text{ for all } r \in A_\Psi .$$

$$\int_s^r \left( \int_{dX(t)} \Omega^{m-1} \wedge d^c h \right) u(t)\,dt$$

$$= \int_s^r \int_{dX(t)} j_t^*(F_2 - F_3)\,u(t)\,dt$$

$$= \int_{X(s,r)} d\Psi \wedge \Psi * u\,(F_2 - F_3) \qquad \text{(by Fubini's theorem)}$$

$$= \int_{X(s,r)} d(\Psi * U) \wedge \Omega^{m-1} \wedge d^c h \qquad (U(r) = \|u\|_{[L^I 1,r]})$$

$$= \int_{X(s,r)} d(h\Omega^{m-1}) \wedge d^c(\Psi * U) .$$

On the other hand, (1.8) and the Stokes theorem ([10, 7.1.3]) imply that

$$M_h(r) - M_h(a) - \int_{X(s,r)} h\Omega^m = \int_{X(s,r)} dh \wedge \Sigma m.$$

Hence by (1.3) the conclusion (1.9) follows:

To prove (1.10), let $w \in (X_*)_{reg}$ with $d\Psi(w) \neq 0$.  Then one can choose a local patch $\alpha: W \longrightarrow V \subseteq C^m$ of $X$ at $w$ and an orientation preserving diffeo-

238                                                   C. TUNG

morphism $\beta: V \longrightarrow V' \times V''$ such that the following diagram commutes:

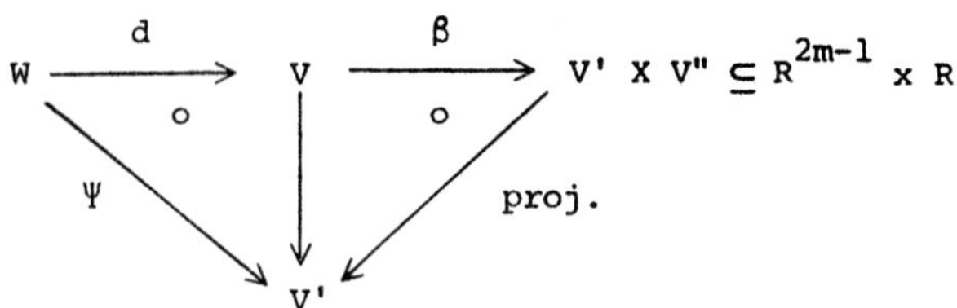

In $W$ the form $\Omega^{m-1} \wedge d^C h$ can be written

$$\Omega^{m-1} \wedge d^C h = \xi \wedge (d\Psi|W) + F$$

with

$$F = k \, \rho^*(dr_1 \wedge \ldots \wedge dr_{2m-1}),$$

where $\xi$ is a continuous form of degree $(2m-2)$ on $W$, $k \in C^0(W)$, $\rho = \beta \circ d$, and $r_1, \ldots, r_{2m}$ are the natural coordinates on $R^{2m}$. Let $z = (z_1, \ldots, z_m) \in V$ and $(a,t) = \beta(z) \in V' \times V''$. There exist functions $x_j \in C^\infty(V' \times V'')$, $1 \leq j \leq 2m$, such that

$$z_j = (x_{2j-1} + i \, x_{2j}) \circ \beta \qquad \text{on } V \text{ for } j = 1, \ldots, m.$$

For each $t \in V''$ set

$$g_t^\alpha(s) = \left( \sum_{j=1}^{2m} \cdot \left| g_{t',j}^\alpha(s) \right|^2 \right)^{\frac{1}{2}} \qquad\qquad (s \in V')$$

where

$$g_{t,j}^\alpha(s) = \frac{\partial(x_1, \ldots, x_{j-1}, x_{j+1}, \ldots, x_{2m})}{\partial(r_1, \ldots \ldots \ldots, r_{2m})} (s,t).$$

A volume element of $W_t = W \cap \Psi^{-1}(t)$ is given by

$$U_t^\alpha = \rho_t^*(g_t^\alpha \, dr_1 \wedge \ldots \wedge dr_{2m-1}) > 0,$$

where $\rho_t = \rho|W_t$. Then there exist $Q_\mu^\alpha \in C^0(W)$ ($\mu = 2,3$) such that for each $t \in A_\Psi \cap V''$,

$$j_{t,}^*(F) = j_t^*(Q_2^\alpha - Q_3^\alpha) \, U_t^\alpha \qquad \text{on } W_t$$

with $j_t^*(Q_\mu^\alpha) \geq 0$ for $\mu = 2,3$. Consequently the assertion (1.0) follows.
Q.E.D.

A function $h : X \longrightarrow C$ is said to be : (1) of type A relative to $\Psi$ iff $h \in c^0(X) \cap c^1(X_*)$ with $h \mid \Psi^{-1}(0) = $ constant; (2) of type B relative to $\Psi$ iff $h \in c^2(X)$ and $\partial\bar{\partial}h \wedge \Theta^{m-1} = 0$ in $X_{reg}$. Obviously every $c^2$-weakly holomorphic (or $c^2$-pluriharmonic) function on $X$ is of type B relative to $\Psi$. Let $R_q$ denote any one of the forms $\Sigma_q$, $B_q$, $\bar{B}_q$.

Lemma 1.2. Let $h : X \longrightarrow C$ be of type B relative to $\Psi$ and $Q : R(0,\infty) \longrightarrow C$ of class $c^1$. Then setting $Q_\Psi = Q \circ \Psi$,

$$\int_{X(s,r)} \Theta^{m-1} \wedge d^c h_\Psi \wedge dh = \int_{X(s,r)} \Theta^{m-1} \wedge \bar{\partial}Q_\Psi \wedge dh = 0, \tag{1.11}$$

$$\int_{X(s,r)} R_m \wedge dh = 0 \tag{1.12}$$

for all $r$, $s$ with $0 < s < r < b$.

Proof. In the manifold $X_{reg}$ one has

$$-\Theta^{m-1} \wedge d^c Q_\Psi \wedge dh = \Theta^{m-1} \wedge dQ_\Psi \wedge d^c h.$$

Thus for all $r$, $s$ in $A_\Psi$ with $s < r$,

$$\int_{X(s,r)} Q_\Psi \Theta^{m-1} \wedge dd^c h - d^c Q_\Psi \wedge dh \wedge \Theta^{m-1}$$

$$= \int_{X(s,r)} d(Q_\Psi d^c h \wedge \Theta^{m-1})$$

$$= Q(t) \int_{dX(t)} \Theta^{m-1} \wedge d^c h \Bigg]_{t=s}^{t=r}$$

$$= 0.$$

In view of the denseness of $A_\Psi$ in $R(0,b)$, the first vanishing assertion in (1.11) follows. Similarly one obtains the second half of (1.11).

240                               C. TUNG

Using (1.1)-(1.4) the relation (1.12) is easily derived from (1.11). Q.E.D.

Remark. Assume $\Psi$ is <u>central</u>, i.e., the set $X[0]$ has measure zero. Then with h as above and $X(r)$ in place of $X(s,r)$, the assertions (1.11)-(1.12) remain valid, provided the integrals over $X(r)$ are defined as limiting values of corresponding integrals over $X(s,r)$ when $s \longrightarrow 0$.

For $k \in C^0(X)$ define

$$L_k(r) = u(r)^m \int_{X(r)} k \, \theta^m \qquad (0 < r < b).$$

By (1.8) and (1.5) one has

$$|M_k(r)| \leq \sup \|k\| \, L_1(r) \qquad (r \in A_\Psi). \tag{1.13}$$

If $k : X \longrightarrow C$ is of type B relative to $\Psi$, it follows from (1.2)-(1.4) and Lemma 1.2 that

$$L_k(r) = \int_{dX(r)} k \, R_m \qquad (r \in A_\Psi). \tag{1.14}$$

Assume $\Psi$ is central. By (1.5) and the Stokes theorem, the function $M_1(r)$ is non-negative and increasing on A. Consequently if $k \in C^0(X)$, the form $k \, \Omega^m$ is integrable on $X(r)$ for each $r \in R(0,b)$. Also, if $k \in C^2(X)$ is of type B relative to $\Psi$, then by (1.12) one has

$$M_k(r) = M_k(0+) + \int_{X(r)} k\Omega^m \qquad (r \in A_\Psi). \tag{1.15}$$

Theorem 1.3. Assume $\Psi$ is central and real analytic. If $h \in C^0(X)$ is of type A relative to $\Psi$, then for each $z \in \Psi^{-1}(0)$,

$$M_1(0+) \, h(z) = M_h(r) - \int_{X(r)} h \, \Omega^m - \int_{0+}^{r} P_h(t) \, u(t) dt. \tag{1.16}$$

Proof. if $z \in \Psi^{-1}(0)$ and $k = h - h(0)$, it follows from (1.13)-(1.14) that $M_k(r) \longrightarrow 0$ as $r \longrightarrow 0$ in $A_\Psi$. Since $M_h = M_k + h(z) M_1$, letting $s \longrightarrow 0$

in (1.9) one obtains the relation (1.16).   Q.E.D.

## 2.  Semi-analytic covering spaces

Let X, Y be complex spaces.  A holomorphic map $\pi : X \longrightarrow Y$ is called a semi-analytic covering of Y iff $\pi$ is surjective, proper, and there exists a thin anlytic subset S of Y such that the set $T = \pi^{-1}(S)$ is thin in X and the restriction $\pi \mid X - T$ defines an analytic covering of $Y_0 : = Y - S$ (the associated sheet numbering being denoted by $s(\pi)$).

Lemma 2.1.   Assume $\pi : X \longrightarrow Y$ is a surjective, holomorphic map of strick rank m([1, p. 17]), where m = dim X.   Then Y is pure m-dimensional.

Proof.   Let $E(\pi) = \{x \in X \mid \text{rank}_x \pi < m\}$, and B be a branch of Y.   Since $E(\pi)$ is thin in X, $\pi^{-1}(B \cap Y_{\text{reg}}) \not\subseteq E(\pi)$.   Hence it follows from [1,1.21] that B has dimension m.   Consequently Y is pure m-dimensional.   Q.E.D.

Let $f : X \longrightarrow D$ be a holomorphic map of X into an open set D in $C^n$. For each $z \in X$ set

$$\psi^z = \|f - f(z)\|^2$$

$$\Theta^{(z)} = dd^c \psi^z.$$

As in § 1, one may define the singular forms $\Omega^{(z)}$, $\Sigma^{(z)}$, etc., in terms of $\psi = \psi^z$, and g = log.  Using the euclidean metric in $C^n$ define $a(z) = \text{dist}(f(z), \partial D)$.   Set $I_z = R(0, a(z)^2)$, $X^z(r) = \{\psi^z < r\}$ for r > 0.

Now assume $f : X \longrightarrow$ is proper.   Let $z \in X$ and $U = X^z(a(z)^2)$.   Define

$$A_f^z(r) = r^{-2m} \int_{U(r^2)} \Theta^{(z),m} \qquad (0 < r < a(z)). \qquad (2.1)$$

If $h \in C^0(U)$ is of type A relative to $\psi^z \mid U$, the relation (1.16) implies that the integral

$$c_f^z(h) = \int_{dU(r)} h\sum{}_m^{(z)} - \int_{U(r)} h\Omega^{(z),m} - \int_{0+}^{r} \left( \int_{dU(t)} \Omega^{(z),m-1} \wedge d^c h \right) \frac{dt}{t} \qquad (2.2)$$

is independent of $r \in I_z$; moreover, by (1.14) and (2.1),

$$c_f^z(h) = h(z) \lim_{r \to 0} A_f^z(r).$$ 
(2.3)

Observe that if h is of type B relative to $\psi^z|U$, by virtue of (1.9), (1.12) and (1.15) the constancy of $c_f^z(h)$ remains valid, the last integral in (2.2) being equal to 0.

Let $p : W \longrightarrow W'$ be a local patch of X at z, where $W' = p(W)$ is analytic in an open subset of $C^n$. Then by (2.3), the constant $c_p^z(1)$ yields the Lelong number of W' at $z' = p(z)$. According to Draper [2] the Lelong number agrees with the algebraic multiplicity of the local ring of W' at z'. Thus for every point z of X a positive integer

$$L_X(z) := c_p^z(1)$$

may be assigned. This leads to a characterization of simple points of X (Corollary 2.3). At first a Cauchy type formula shall be derived:

Theorem 2.2. Assume $f : X \longrightarrow D$ defines a semi-analytic covering of $Y \subseteq D$. Let $z \in X$ and $V = X^{(z)}(b)$ with $b \in I_z$. If $h \in C^0(V)$ is of type A relative to $\psi^z|V$, then

$$s(f) \, L_Y(z') \, h(z) = c_f^z(h)$$ 
(2.4)

where $z' = f(z)$.

Proof. Let j denote the inclusion : $Y \longrightarrow C^{n,z'}(b^{1/2})$. By (2.1) and [10, 5.2.2], one has

$$A_f^z(r) = s(f) \, A_j^{z'}(r) \qquad (0 < r < b^{1/2}).$$ 
(2.5)

Hence (2.3) implies that (2.4) holds. Q.E.D.

Corollary 2.3. A point z of a complex space X is non-singular if there exist (i) an injective, proper, holomorphic map f of a neighborhood w of z into an open set $D \subseteq C^n$; (ii) a function $h \in C^2(V)$, where $V = W^z(b)$ for some $b \in I_z$, such that h is of type B relative to $\psi^z|V$ and

$$0 \neq h(z) = \int_{dV(r)} h_m^{(z)} - \int_{V(r)} h\Omega^{(z)}, m$$ 
(2.6)

for some $r \in R(o,b)$.

Proof.  If $z \in X_{reg}$, there exists a local patch $f : W \longrightarrow W' \subset \mathbb{C}^m$ of X at z.  Then by (2.4) the assertion (ii) holds with $h = 1$.

Now assume that there exist f and h satisfying the conditions (i)-(ii). With $Y = f(W)$, (2.4) and (2.6) yield

$$L_Y(z') = C_f^z(h)/h(z) = 1.$$

Thus $z' \in W'_{reg}$.  It follows that f is biholomorphic in a neighborhood of z. Q.E.D.


### 3.  The main result

Proof of Theorem I.  **Sufficiency**.  Let f and Q be given with properties (a)-(b).  By Lemma 2.1, the space $Y = f(X)$ is pure m-dimensional.  Since Y is irreducible ([1, 1.17]), the map f induces a semi-analytic covering of Y ([1, 2.2]) with sheet number s(f).

If $\Psi^{-1}(0) \neq \emptyset$, with $h = \Psi*Q$ (1.16) yields

$$Q(r)\, M_1(r) = C + \int_{X(r)} (\Psi*Q)\, \Omega^m + \int_0^r \left( \int_{dX(t)} \Omega^{m-1} \wedge d^c\Psi \right) \frac{Q'(t)}{t}\, dt \qquad (3.1)$$

for all $r > 0$, where C is a constant $\geq 0$.  If $\Psi^{-1}(0) = \emptyset$, by (1.9) the above relation remains valid with $C = 0$.  Hence (0.1), (1.40 and (3.1) imply that

$$M_1(r) = O(1) \qquad (r \longrightarrow \infty). \qquad (3.2)$$

Let $j : Y \longrightarrow \mathbb{C}^n$ be the inclusion.  As in (2.5) one has

$$L_1(r^2) = s(f)\, A_j^0(r) \qquad (r > 0).$$

Now (3.2) and (1.14) yield

$$A_j^0(r) = O(1) \qquad (r \longrightarrow \infty).$$

According to Stoll [7,II] Y is an algebraic variety in $\mathbb{C}^n$.  Hence by Griffiths-King [5, 2.4] there exists a surjective, proper, holomorphic $\beta : Y \longrightarrow \mathbb{C}^m$.  Then $f* \beta : X \longrightarrow \mathbb{C}^m$ defines an analytic cover of $\mathbb{C}^m$.

**Necessity**.  Suppose $\pi: X \longrightarrow \mathbb{C}^m$ defines an analytic cover of $\mathbb{C}^m$.  Then X is of type E and $\Psi := \|\pi\|^2$ is an exhaustion function of X with $\Omega^m = 0$. Hence (0.1) holds with $Q = 1$.  Q.E.D.


The preceding proof also yields the following:

Proposition 3.1.  Let X be an irreducible compelx space of dimension m. Assume X admits a proper, holomorphic map f into some $C^n$ with rank f = m, and there exists a continuous function $Q : R [0,\infty) \longrightarrow R$ satisfying the conditions (a)-(b) in Theorem I.  Then X is a semi-analytic covering of $C^m$ (hence is parabolic in the sense of [9]).

## References


1. Andreotti, A., and Stoll, W., Analytic and algebraic dependence of meromorphic functions, Lecture Notes in Mathematics 234.  Springer-Verlag, Berlin and New York, 1971.

2. Draper, R., Intersection theory in analytic geometry, Math. Ann. 180 (1969), 175-204.

3. Grauert, H., On Levi's problem and the imbedding of real-analytic manifolds, Ann. of Math. 68 (1958), 460-472.

4. Grauert, H., and Remmert, R., Theorie der Steinschen Raume, Grundl. Math. Wiss, Band 227, Springer-Verlag, Berlin-New York, 1977.

5. Griffiths, P., and King, J., Nevanlinna theory and holomorphic mappings between algebraic varieties, Acta Math. 130 (1973), 145-220.

6. Narasimhan, R., Imbedding of holomorphically complete complex spaces, Amer. J. Math. 82 (1960), 917-934.

7. Stoll, W., The growth of the area of a transcendental analytic set I, Math. Ann 156 (1964), 47-78, II, Math. Ann. 156 (1964), 144-170.

8.       , Value distribution of holomorphic maps into compact complex manifolds, Lecture Notes in Mathematics 135.  Springer-Verlag, Berlin-Heidelberg-New York, 1970.

9.       , Value distribution on parabolic spaces, Lecture Notes in Mathematics 600.  Springer-Verlag, Berlin-Heidelberg-New York, 1977.

10. Tung, C., The first main theorem of value distribution on complex spaces, Memorie dell'Accademia Nazionale dei Lincei, Serie VIII, Vol. XV, Sez. 1, Fasc. (1979), 91-263.

11.       , Equidistribution theory in higher dimensions, Pacific J. Math. 78 (1978), 525-548.



Department of Mathematics, Astron. & Statistics
Mankato State University
Mankato, MN  56001

Contemporary Mathematics
Volume 25, 1983

MEROMORPHIC SOLUTIONS OF SOME DIFFERENCE EQUATIONS

AND THE VALUE DISTRIBUTION THEORY IN HALF-STRIPS

by

Niro Yanagihara

## 1. Introduction

Here we consider the difference equation

$$(1.1) \qquad \alpha_n w(z+n) + \alpha_{n-1} w(z+n-1) + \ldots + \alpha_1 w(z+1) = R(w(z)),$$

where $R(w)$ is a rational function of $w$:

$$R(w) = P(s)/Q(w),$$
$$(1.2) \qquad P(w) = a_p w^p + \ldots + a_0,$$
$$Q(w) = b_q w^q + \ldots + b_0,$$

in which $\alpha_n, \ldots, \alpha_1; a_p, \ldots, a_0; b_q, \ldots, b_0$ are constants, $\alpha_n a_p b_q \neq 0$. $P(w)$ and $Q(w)$ are supposed to be mutually prime. Below, we denote by $p$ and $q$ the degree of the nominator $P(w)$ and the denominator $Q(w)$, respectively. Further, we assume that $n > 1$.

Denote

$$(1.3) \qquad q_0 = \max(p, q).$$

we proved in [6] and [7]

PROPOSITION 1. Suppose $p \geq q + 2$. Then any meromorphic solution $w(z)$ of (1.1) is transcendental and of order $\infty$ in the sense of Nevanlinna.

PROPOSITION 2. Suppose $q_0 \geq n + 1$. Then any meromorphic solution $w(z)$ of (1.1) is transcendental and of order $\infty$ in the sense of Nevanlinna.

PROPOSITION 3. Suppose $p = q + 1$. Then any meromorphic solution $w(z)$ of (1.1) is transcendental. If $q_0 \leq n$, then there is an equation of the form (1.1) which admits a solution of finite order in the sense of Nevanlinna.

PROPOSITION 4. Suppose $p \leq q$ and $q_0 \leq n$. Then there is an equation of the form (1.1) which admits a rational solution.

Put

$$(1.4) \qquad \Lambda = \left\{ \lambda; \ |\lambda| < \infty, \ (\alpha_n + \ldots + \alpha_1)\lambda = R(\lambda) \right\}.$$

We proved in [8]

PROPOSITION 5. Suppose $p \geq q + 2$. Let $w(z)$ be a meromorphic solution of (1.1). Then there is a $\lambda \in \Lambda$ such that

$$(1.5) \qquad w(z - \mu) \to \lambda \text{ as } \mu \uparrow \infty.$$

In view of Propositions 1, 2, 3, one may conjecture that (1.5) would hold for the case $p = q + 1 = q_0 \geq n + 1$. Our aim is to prove this in a modified form. To the purpose, put

(1.6)     $\Lambda_0 = \Lambda \cup \infty$ .

Then

THEOREM 6.  Suppose $p = q + 1 = q_0 \geq n + 1$.  Further suppose that $Q(w)$ in (1.2) has at least two zeros.  Let $w(z)$ be a meromorphic solution of (1.1).  Then there is a $\lambda \in \Lambda_0$ ($|\lambda| \leq \infty$) for which (1.5) holds.

For the proof of Theorem 6, we need the Nevanlinna theory in a half-strip.

2.  Nevanlinna theory in a half-strip domain

For a positive number $A > 0$ and real numbers $a$, $a'$ ($a < a'$), put

(2.1)     $H = H(A, a, a') = \left\{ z = x+iy; \ -A < y < A, \ a < x < a' \right\}$ .

Let sn be the Jacobian elliptic function with the fundamental periods $4A$ and $2iA'$, $A > 0$, $A' > 0$.  It is well known that $sn(iz-ia)$ maps $H(A, a, a+A')$ onto the upper half-plane.  Write

(2.2)     $v(z) = v(z; A, a, a', c) =$

$= \log \left| [sn(iz-ia) + sn(ic-ia)]/[sn(iz-ia) - sn(ic-ia)] \right|$ ,

where $c$ is a fixed constant, $a < c < a' = a+A'$.  $v(z)$ is the Green's function of $H(A, a, a')$ with the pole at $z = c$.

Let $f(z)$ be a function meromorphic on $\overline{H(A, a, \infty)}$.  Write

(2.3)     $m(A, a, a', c' \infty, f) = \dfrac{1}{2\pi} \int_{\partial H} \dfrac{\partial v}{\partial n} \log \dfrac{\sqrt{1 + |f(z)|^2}}{\sqrt{1 + |f(c)|^2}} \, ds_z$ ,

$N(A, a, a', c; \infty, f) = \Sigma \, v(z_n)$ ,

where summation is made with respect to poles $z_n$ of $f(z)$ in $H(A, a, a')$.

If $c$ is a pole of order $\kappa$ for $f(z)$, put

$m(A, a, a', c; \infty, f) = \dfrac{1}{2\pi} \int_{\partial H} \dfrac{\partial v}{\partial n} \log \dfrac{\sqrt{1 + |f(z)|^2}}{|c_\kappa|} \, ds_z$ ,

(2.3')

$N(A, a, a', c; \infty, f) = \Sigma \, v(z_n) + \kappa V_0$ ,

where $c_\kappa = \lim\limits_{z \to c} [(z - c)^\kappa f(z)]$ and $V_0 = \lim\limits_{z \to c} (v(z) - \log \dfrac{1}{|z - c|})$, and summation is made with respect to poles $z_n \neq c$ of $f(z)$ in $H(A, a, a')$.  Further put

(2.4)     $T(A, a, a', c; f) = \dfrac{1}{\pi} \iint_H v(z) \left( \dfrac{|f'(z)|^2}{1 + |f(z)|^2} \right) d\sigma_z$ ,

where $d\sigma_z$ is the surface element.

Write, for a complex number $b$,

$m(A, a, a', c; b, f) = m(A, a, a', c; \infty, (1 + \overline{b}f)/(f - b))$,

$N(A, a, a', c; b, f) = N(A, a, a', c; \infty, (1 + \overline{b}f)/(f - b))$.

Applying the Stokes formula to $v(z)$ and $u(z) = \log \sqrt{1 + |f(z)|^2}$, we obtain the first fundamental theorem in the half-strip:

THEOREM 2.1.  For any $b$, $|b| \leq \infty$,

$$m(A, a, a', c; b, f) + N(A, a, a', c; b, f) =$$

$$T(A, a, a', c; f) = T(A, a, a', c; 1/f) .$$

We will show the following theorem:

THEOREM 2.2. Let $R(w) = P(w)/Q(w)$ be a rational function of degree $q_0$. If $f(z)$ is meromorphic on $H(A, a, \infty)$, we have

$$(2.5) \qquad T(A, a, a', c; R(f)) = q_0 T(A, a, a', c; f) + 0(1).$$

$0(1)$ is dependent on $f(c)$, and independent of $A, a, a'$, if the point $c(a < c < a')$ is fixed.

REMARK. By Theorem 2.1, we can suppose that $p > q$ without losing generality.

Proof. Let $\Gamma$ be the union of a family of closed disks (finite in number) which contain all zeros of $P(w)$ and do not contain any zero of $Q(w)$. There are constants $M_1$ and $M_2$ such that

$$(2.6) \qquad M_1/|P(w)| \le |1/R(w)| \le M_2/|P(w)| \quad \text{for } w \in \Gamma .$$

$1/R(w)$ as well as $1/P(w)$ is bounded outside $\Gamma$, i.e.,

$$(2.6') \qquad |1/R(w)| \le L, \quad |1/P(w)| \le L \quad \text{if } w \notin \Gamma$$

for a constant L. Let

$$E_1 = \{z; z \in \partial H(A, a, a'), f(z) \in \Gamma\},$$

$$E_2 = \{z; z \in \partial H(A, a, a'), f(z) \notin \Gamma\}.$$

Then, since

$$\left| \int_{E_2} \frac{\partial v}{\partial n} \log[\sqrt{1 + |1/R(f(z))|}^2 /\sqrt{1 + |1/R(f(c))|}^2] \, ds_z \right| \le$$

$$\le 2\pi \max(\log\sqrt{1 + L^2}, \log\sqrt{1 + |1/R(f(c))|}^2),$$

we obtain

$$(2.7) \qquad m(A, a, a', c; \infty, \frac{1}{R(f)}) = \frac{1}{2\pi}\int_{E_1} + \frac{1}{2\pi}\int_{E_2} = \frac{1}{2\pi}\int_{E_1} + 0(1),$$

where $0(1)$ is independent of $A, a, a'$. Similarly

$$\left| \int_{E_2} \frac{\partial v}{\partial n} \log[\sqrt{1 + |1/P(f(z))|}^2 /\sqrt{1 + |1/P(f(c))|}^2] \, ds_z \right| \le$$

$$\le 2\pi \max( \log\sqrt{1 + L^2}, \log\sqrt{1 + |1/P(f(c))|}^2 ,$$

and

$$(2.7') \qquad m(A, a, a', c; \infty, 1/P(f)) = \frac{1}{2\pi}\int_{E_1} + 0(1) ,$$

where $0(1)$ is independent of $A, a, a'$.

Therefore, by (2.6), (2.7), and (2.7'), we get

$$m(A, a, a', c; \infty, 1/R(f)) = m(A, a, a', c; \infty, 1/P(f)) + 0(1),$$

where $0(1)$ is independent of $A, a, a'$. Since

$$N(A, a, a', c; \infty, 1/R(f)) = N(A, a, a', c; \infty, 1/P(f)),$$

N. YANAGIHARA



we obtain

$$(2.8) \quad T(A, a, a', c; 1/R(f)) = T(A, a, a', c; 1/P(f)) + 0(1) =$$
$$= T(A, a, a', c; P(f)) + 0(1).$$

Let C* be a number such that, if $P(w) = a_p w^p + a_{p-1} w^{p-1} + \ldots$,

$$2|a_p w^p| \geq |P(w)| \geq (1/2)|a_p w^p| \quad \text{if } |w| \geq C* ,$$

and put

$$E_1^* = \{z; z \in \partial H(A, a, a'), |f(z)| > C*\} ,$$

$$E_2^* = \{z; z \in \partial H(A, a, a'), |f(z)| \leq C*\} .$$

As above, we obtain

$$(2.9) \quad m(A, a, a', c; \infty, P(f)) = \frac{1}{2\pi} \int_{E_1^*} \frac{\partial v}{\partial n} \log \frac{\sqrt{1 + |P(f(z))|^2}}{\sqrt{1 + |P(f(c))|^2}} \, ds_z$$

$$+ \frac{1}{2\pi} \int_{E_2^*} = \frac{1}{2\pi} \int_{E_1^*} + 0(1) = q_0 m(A, a, a', c; \infty, f) + 0(1)$$

$(q_0 = p)$, and

$$N(A, a, a', c; \infty, P(f)) = q_0 N(A, a, a', c; \infty, f),$$

hence

$$T(A, a, a', c; R(f)) = q_0 T(A, a, a', c; f) + 0(1) .$$

Since $0(1)$ in (2.8) and (2.9) are independent of A, a, a', we obtain the results.     Q.E.D.

THEOREM 2.3. <u>Let</u> $f_1(z)$ <u>and</u> $f_2(z)$ <u>be meromorphic on</u> $\overline{H(A, a, \infty)}$.
<u>Then</u>

$$(2.10) \quad T(A, a, a', c; f_1 + f_2) \leq T(A, a, a', c; f_1) + T(A, a, a', c; f_2) + 0(1),$$

$$(2.10') \quad T(A, a, a', c; f_1 f_2) \leq T(A, a, a', c; f_1) + T(A, a, a', c; f_2) + 0(1),$$

<u>where</u> $0(1)$ <u>is dependent on</u> $f_1(c)$ <u>and</u> $f_2(c)$, <u>and independent of</u> A, a, a', <u>if</u> c $(a < c < a')$ <u>is fixed.</u>

<u>Proof</u> is obvious from the inequalities

$$\sqrt{1 + |f_1(z) + f_2(z)|^2} \leq \sqrt{2} \sqrt{1 + |f_1(z)|^2} \sqrt{1 + |f_2(z)|^2} ,$$

$$\sqrt{1 + |f_1(z) f_2(z)|^2} \leq \sqrt{1 + |f_1(z)|^2} \sqrt{1 + |f_2(z)|^2} .$$

THEOREM 2.4. <u>Let</u> $A \leq A_1$ , $a \geq a_1$, $a' \leq a_1'$, <u>then</u>

$$T(A, a, a', c; f) \leq T(A_1, a_1, a_1', c; f) .$$

The proof is obvious.

3. <u>Preliminary lemmas</u>

Let $w(z)$ be a meromorphic solution of (1.1), in which we suppose that $p = q + 1 \geq n + 1$.

For a constant K, we put

(3.1)        $L(y_0, K) = \{z = x + iy_0;\ x \leq K\}$ ,

and

$$\Delta = \{y_0;\ w(z) \text{ has poles on } L(y_0, K) \text{ for any } K\}\ ,$$
(3.2)

$$H^*(\alpha, \beta, K) = \{z = x+iy;\ \alpha < y < \beta,\ x \leq K\}\ .$$

LEMMA 3.1.  <u>Let</u> $y_0 \in \Delta$. <u>There are</u> $\alpha, \beta\ (\alpha < y_0 < \beta)$, <u>and K such that</u> $w(z)$ <u>is holomorphic on</u> $H^*(\alpha, y_0, K) \cup H^*(y_0, \beta, K)$.

<u>Proof.</u>  Suppose there are poles $\{z_m\}$ of $w(z)$ such that $z_m = x_m + iy_m$, $y_m \to y_0$, $x_m \to -\infty$. For each m, there are integers $j_1, \ldots, j_{k_m}$ such that $1 \leq j_\ell \leq n\ (1 \leq \ell \leq k_m)$, $-n \leq x_m + j_1 + j_2 + \ldots + j_{k_m} \leq 0$, and $z_m + j_1 + \ldots + j_\ell\ (1 \leq \ell \leq k_m)$ are poles of $w(z)$. Put $z'_m = z_m + j_1 + \ldots + j_{k_m} = x'_m + iy_m$. Then poles $\{z'_m\}$ clusters at a point $z_0 = x_0 + iy_0$, $-n \leq x_0 \leq 0$, which is a contradiction.          Q.E.D.

LEMMA 3.2.  <u>Let</u> $R(w) = P(w)/Q(w)$ <u>be the rational function in</u> (1.2), $p = q + 1$, <u>and</u> $w_1$ <u>be a zero of</u> $Q(w)$. <u>Let</u> $y_0 \in \Delta$ <u>and</u> $\alpha, \beta, K$ <u>be the numbers in</u> Lemma 3.1.  <u>There is a number</u> $K' < K$ <u>such that</u> $w(z)$ <u>does not take</u> $w_1$ <u>in</u> $H^*(\alpha, y_0, K') \cup H^*(y_0, \beta, K')$.

<u>Proof.</u>  Suppose there is a sequence $z_m$, $z_m = x_m + iy_m$, such that $y_m \to y_0$, $x_m \to -\infty$, and $w(z_m) = w_1$. Then there is a pole $z_m^* = z_m + j^{(m)}$, $1 \leq j^{(m)} \leq n$, of $w(z)$, which contradicts with Lemma 3.1.      Q.E.D.

LEMMA 3.3.  <u>Let</u> $y_0 \in \Delta$. <u>There are a number</u> K <u>and poles</u> $z_1, \ldots, z_k$, $z_j = x_j + iy_0$, $K - n \leq x_j \leq K$, <u>such that, if</u> $z_0$ <u>is a pole on</u> $L(y_0, K)$, <u>then</u> $z_j - z_0$ <u>is an integer and</u> $\text{ord}(z_j) = \text{ord}(z_0)$ <u>for a</u> j, $1 \leq j \leq k$.

LEMMA 3.4.  <u>Let</u> $w_1$ <u>be a zero of</u> $Q(w)$ <u>in</u> (1.2).  <u>Let</u> $y_0 \in \Delta$. <u>There are a number</u> $K'$ <u>and</u> $w_1$<u>-points</u> $z'_1, \ldots, z'_k$, $z'_j = x'_j + iy_0$, $K' - n \leq x'_j \leq K'$ <u>such that, if</u> $z'_0 \in L(y_0, K')$ <u>and</u> $w(z'_0) = w_1$, <u>then</u> $z'_j - z'_0$ <u>is an integer for a</u> j, $1 \leq j \leq k'$ .

Proofs of Lemmas 3.3 and 3.4 are easy and may be omitted.

4.  <u>Preliminary inequalities</u>

Put, for a meromorphic function $f(z)$,

$$S(A, a, a'; f) = \frac{1}{\pi} \int_{-A}^{A} dy \int_{a}^{a'} \left( \frac{|f'(\xi+iy)|}{1 + |f(\xi+iy)|^2} \right)^2 d\xi\ .$$

$S(A, a, a'; f)$ is the spherical area of the image of $H(A, a, a')$ by the mapping $f(z)$.  Let $b_1, b_2, b_3$ be three distinct values.  Then by [5, p. 100, Lemma 2], we have

$$(4.1) \qquad S(A, a, a'; f) \leq 3 \sum_{j=1}^{3} n(a' + 2 + A;\ b_j, f) + 0(a'),$$

N. YANAGIHARA

where $n(t; b_j, f)$ is the number of zeros of $f(z) - b_j$ in $H(A, a, t)$, multiple zeros being counted only once.

It is obvious that $\frac{\partial v}{\partial x}(z; A, a, a', c)$ is bounded if $x = \text{Re } z \geq c+1$, as $a' \to \infty$. Hence we have

$$T(A, a, a', c; f) = \frac{1}{\pi} \int_{-A}^{A} dy \int_{a}^{a'} v(x+iy) \left(\frac{|f'(x+iy)|}{1 + |f(x+iy)|^2}\right)^2 dx .$$

$$= \frac{1}{\pi} \int_{-A}^{A} dy \int_{c+1}^{a'} -[\frac{\partial v}{\partial x}(x+iy) \int_{c+1}^{x} \left(\frac{|f'(\xi+iy)|}{1 + |f(\xi+iy)|^2}\right)^2 d\xi] dx + O(1)$$

$$\leq K \int_{a}^{a'} S(A, a, x; f) dx + O(1)$$

with a constant $K > 0$. Thus by (4.1), we have

$$(4.2) \qquad T(A, a, a', c; f) \leq 3K \sum_{j=1}^{3} \int_{a}^{a'} n(x+2+A; b_j, f) dx + O(a'^2) .$$

On the other hand, it is easy to see that, for any $\varepsilon > 0$, we have

$$(4.3) \qquad v(z; A, a, a', c+1) \leq (1 + \varepsilon) v(z; A, a, a', c)$$

supposed that $x = \text{Re } z$ and $a'$ are sufficiently large.

Let $w(z)$ be a meromorphic solution of (1.1). Then by Theorems 2.2 and 2.3,

$$T(A, a, a', c; R(w(z))) = q_0 T(A, a, a', c; w(z)) + O(1)$$

$$= T(A, a, a', c; \alpha_n w(z+n) + \ldots + \alpha_1 w(z+1)$$

$$\leq \sum_{k=1}^{n} T(A, a, a', c; w(z+k)) + O(1) =$$

$$= \sum_{k=1}^{n} T(A, a+k, a'+k, c+k; w(z)) + O(1)$$

$$\leq n(1 + \varepsilon) T(A, a, a'+n, c; w(z)) + O(1)$$

if $a'$ is sufficiently large, where $O(1)$ is determined by $\varepsilon > 0$. Thus

$$(4.4) \qquad q_0 T(A, a, a', c; w(z)) \leq n(1+\varepsilon) T(A, a, a'+n, c; w(z)) + O(1)$$

If $T(A, a, a', c; w(z)) \to \infty$ as $a' \to \infty$, then we can say that

$$q_0 T(A, a, a', c; w(z)) \leq n(1 + 2\varepsilon) T(A, a, a'+n, c; w(z)).$$

Suppose $q_0 \geq n+1$. If $\varepsilon$ is so small that $q_1 = q_0/[n(1 + 2\varepsilon)] > 1$, then

$$T(A, a, a'+n, c; w(z)) \geq q_1 T(A, a, a', c; w(z)),$$

and

$$T(A, a, a'+mn, c; w(z)) \geq q_1^m T(A, a, a', c; w(z)).$$

Therefore

$$(4.5) \qquad T(A, a, a', c; w(z)) \geq M q_2^{a'} \quad \text{as } a' \to \infty ,$$

with a constant $M > 0$, where $q_2 = q_1^{1/n} > 1$.

5.  Proof of Theorem 6

Now we are ready to prove Theorem 6, at which we have aimed.

Let $w_1, \ldots, w_\ell$ be zeros of $Q(w)$. Then $\ell \geq 2$.

First, suppose that the set $\Delta$ in (3.2) is void. Then, it is easy to see that, for any $\alpha, \beta$, there is a $K$ such that $w(z)$ is holomorphic and does not take any of $w_1, \ldots, w_\ell$ in $H^*(\alpha, \beta, K)$. Hence $\{w(z - \mu)\}_{\mu \geq 0}$ is a normal family, and there is a subsequence $\{\mu_m\}$ such that $w(z - \mu_m)$ converges to a meromorphic solution $W(z)$ of (1.1), uniformly on any compact set. $W(z)$ is holomorphic and does not take any of $w_1, \ldots, w_\ell$, which shows that $W(z)$ is a constant $\varepsilon \Delta_0$ (see (1.6)). Since the cluster set of $w(z)$ is either a continuum or a point, we can conclude that $\{w(z - \mu)\}$ itself converges uniformly on any compact set.

Next suppose that the set $\Delta$ is not void. Take $y_0 \varepsilon \Delta$. By Lemmas 3.3 and 3.4, we obtain $z_1^{(t)}, \ldots, z_{k_t}^{(t)}$, $t = 0, 1, \ldots, \ell$, on $L(y_0, K)$ such that ($K$ is supposed sufficiently large), if we write $z_j^{(t)} = x_j^{(t)} + iy_0$, $K - n \leq x_j^{(t)} \leq K$, $t = 0, 1, \ldots, \ell$, and $z_1^{(0)}, \ldots, z_{k_0}^{(0)}$ are poles stated in Lemma 3.3, and $z_1^{(t)}, \ldots, z_{k_t}^{(t)}$ are $w_t$-points stated in Lemma 3.4, for $t = 1, \ldots, \ell$. Let $r > 0$ be small that $D(z_j^{(t)}, r) = \{|z - z_j^{(t)}| \leq r\}$ are mutually disjoint. For any $\alpha, \beta$, there is a $K'$ such that $w(z)$ does not take any of $\infty, w_1, \ldots, w_\ell$ in

(5.1)    $H^*_{(r)}(\alpha, \beta, K') = H^*(\alpha, \beta, K') \quad \{\bigcup_{j,k,m} D(z_j^{(t)} -m, r) ;$

$$j=1, \ldots, k_t; \ t=0, \ldots,; \ m=0,1,..\}$$

Thus $\{w(z - \mu); \mu \geq 0\}$ is a normal family in $H^*_{(r)}(\alpha, \beta, K')$. Considering $1/w(z - \mu)$ also, $\{w(z - \mu); \mu \geq 0\}$ is seen to be a normal family in $H^*(\alpha, \beta, K')$, hence there is a subsequence $\{w(z - \mu_m); m \geq 1\}$ such that

$$w(z - \mu_m) \to W(z) \quad \text{uniformly on any compact set,}$$

where $W(z)$ is easily seen to be a meromorphic solution of (1.1). If $W(z)$ is not constant, we can take a real number $c$ so that $W(c) \neq \infty$, $\neq$ zeros of $P(w)$, $\neq$ zeros of $Q(w)$.

If $T(A, a, a', c; W(z)) \to \infty$ as $a' \to \infty$, then by (4.5)

(5.2)    $T(A, a, a', c; W(z)) \geq M q_2^{a'}, \ q_2 > 1.$

By Lemmas 3.3 and 3.4, poles and $w_t$-points with constant width on lines $L(y_0, K_{y_0}), \ y_0 \varepsilon \Delta$. Thus $n(a'; w_t, W(z)) = 0(a')$. Hence, by (4.2)

$$T(A, a, a', c; W(z)) \leq 3K \sum_{t=0}^{2} \int_a^{a'} n(x+2+A; w_t, W(z))dx + 0(a'^2)$$

with a constant $K$, where we write $w_0 = \infty$. Thus

$$T(A, a, a', c; W(z)) = 0(a'^2),$$

which contradicts with (5.2). Therefore

(5.3)    $T(A, a, a', c; W(Z))$ is bounded as $a' \to \infty$.

Now

$$T(A, a, a', c; R(W(z))) = q_0 T(A, a, a', c; W(z)) + 0(1)$$

$$= T(A, a, a', c; \alpha_n W(z+n) + \ldots + \alpha_1 W(z+1))$$

$$\leq \sum_{k=1}^{n} T(A, a, a', c; W(z+k)) + 0(1)$$

$$\leq n(1 + \varepsilon) T(A, a, a'+n, c; W(z)) + 0(1),$$

where $\varepsilon > 0$ is chosen so that $n(1 + \varepsilon) < q_0$. We let $a' \to \infty$, then by (5.3), we get

$$q_0 T(A, a, \infty, c; W(z)) \leq n(1 + \varepsilon) T(A, a, \infty, c; W(z)) + 0(1)$$

hence

$$(5.4) \qquad T(A, a, \infty, c; W(z)) \leq M$$

with a constant M, which is independent of A, a, if c is fixed.

Therefore,

$$(5.5) \qquad N(A, a, \infty, c; w_t, W(z)) \leq M$$

for $t = 0, 1, \ldots,$ . Letting $A \to \infty$,

$$v(z; A. a, \infty, c) = \log \left| [\sin(\tfrac{i\pi\tilde{z}}{2A}) + \sin(\tfrac{i\pi\tilde{c}}{2A})] / [\sin(\tfrac{i\pi\tilde{z}}{2A}) - \sin(\tfrac{i\pi\tilde{c}}{2A})] \right|$$

$$\to \log \left| [1 + \tilde{c}/\tilde{z}] / [1 - \tilde{c}/\tilde{z}] \right|, \quad (\tilde{z} = z-a, \ \tilde{c} = c-a)$$

hence (5.5) shows that $w_t$-points of $W(z)$ are finite in number, $t = 0, \ldots, \ell,$ which shows that $W(z)$ is a constant.

Therefore, if $w(z - \mu_m)$ converges, then the limit function is a constant $\varepsilon \ \Lambda_0$. Since the cluster set of $w(z)$ is a continuum or a point, we have that $w(z - \mu)$ itself converges uniformly on any compact set, to a constant $\varepsilon \ \Lambda_0$.

References

1.  A.A. Goldberg and I.V. Ostrovskii: Value Distributions of Meromorphic Functions. Moskva 1970.

2.  R. Nevanlinna: Analytic Functions. Springer-Verlag 1970.

3.  R. Nevanlinna: Über die Eigenschaften meromorpher Funktionen in einem Winkelraum. Acta Soc. Sci. Fenn., tom 50, No. 12 (1925).

4.  M. Tsuji; Potential Theory in Modern Function Theory. Chelsea 1975.

5.  N. Yanagihara: Meromorphic Solution of the difference equation $y(x+1) = y(x) + 1 + \lambda/y(x)$, I. Funkcial. Ekvac., 21 (1978), 97-104.

6.  N. Yanagihara: Meromorphic solutions of some difference equations of higher order, II. Proc. Japan Acad., 58A (1982), 21-24.

7.  N. Yanagihara: Meromorphic solutions of some difference equations of higher order, II. Proc. Japan Acad., 58A (1982), 284-286-

8.  N. Yanagihara: Meromorphic solutions of some difference equations of the n-th order. to appear in Arch. Rat. Mech. Anal.

Department of Mathematics
Faculty of Science
Chiba University
1-33 Yayoi-cho, Chiba City 260
JAPAN